WARMING

WHO IS TAKING THE HEAT?

by
GERALD
FOLEY

PANOS

Published by Panos Publications Ltd
9 White Lion Street
London N1 9PD, UK

A catalogue record for this book is available from the British Library.

Funding for *Global Warming* was provided by the Norwegian Environment Ministry and the Netherlands Environment Ministry.

Any judgements expressed in this document should not be taken to represent the views of Panos or any of its funding agencies.

The Panos Institute is an information and policy studies institute, dedicated to working in partnership with others towards greater public understanding of sustainable development. Panos has offices in Budapest, London, Paris and Washington DC.

For more information about Panos contact:
Juliet Heller, The Panos Institute

Managing editor: Olivia Bennett
Production: Sally O'Leary
Picture research: Adrian Evans
Cover Design: Graphic Partnership
Illustrations: Philip Davies
Printed in Great Britain by J. W. Arrowsmith Ltd., Bristol.

Contents

Preface

The trouble with global warming is that we need to take action before we have proof, because by the time we have incontrovertible evidence of global warming's effects, it will be too late to stop them.

There are some certainties, however. We know that the Earth's atmosphere acts rather like the glass in a greenhouse. It lets the sun's warmth in but stops much of this heat getting out again. Without its atmosphere, the Earth would be as cold as the moon.

We know that certain gases—such as carbon dioxide, water vapour and the man-made chlorofluorocarbons (CFCs)—are what makes the atmosphere work like a greenhouse.

We know that for some decades we have been increasing our emissions of greenhouse gases, and that their concentrations in the atmosphere have been rising. And we are fairly certain that this will increase the Earth's temperature. But by how much? And how fast? And what will be the effects of such warming?

As weather forecasters know, predicting what the atmosphere will do isn't easy. To make it more difficult, the air interacts continually with the oceans, about which we know even less. Water takes a long time to warm up or cool down. So the full warming effect of today's higher concentration of greenhouse gases won't be felt for many decades.

The consensus among scientists is that if we don't do something about it, we can probably expect a worldwide rise of about 1°C by the year 2030 and of 3°C by 2100. That would make our planet hotter than it has been for 120,000 years. On the whole, the effects will be damaging. The sea level will very probably rise, for example. But by how much? And how fast? Has this already started? Again, the scientific estimates are not much more than educated guesses.

What should we do about global warming? We can reduce our reliance on fossil fuels, because burning coal and oil emits carbon dioxide. Adopt measures to promote energy efficiency and penalise over-consumption. Reduce deforestation, because when trees are burned or decay, they emit carbon dioxide. And plant lots of trees, because living trees absorb carbon dioxide. But who should pay?

These and other steps would at best probably only slow down the warming process. We have to plan for at least some rise in global temperatures, and that will cost money. Again, who should pay?

The mainly rich, industrialised nations of the North have been the source of most of the increase in greenhouse gas emissions so far. The mainly poor, developing nations of the South, with generally weaker infrastructures, more fragile environments and fewer resources, would on the whole be more vulnerable to the effects of climate change—and more hurt by cuts in energy use and other steps to reduce the warming.

In the last few years, public opinion in the North has been to some extent alerted to global warming, though there is as yet little agreement on action to counter it. In the South, faced with problems of more immediate concern, what may or may not happen in 40 years is a low priority.

In this Panos book, Gerald Foley sets out the facts on global warming in clear, dispassionate terms. He discusses how both warming and action against it will affect the North and the South. National boundaries will provide no defence against an overheated planet. A global threat requires a global response.

Two important insights emerge from this book. First, almost all the steps needed to counter global warming would bring many other advantages. For example, energy conservation, reforestation and lower levels of atmospheric pollution are all "no regrets" strategies, of benefit to the environment and therefore to humankind whatever the effects of global warming. Second, we shall not be able to take these steps without at the same time tackling hunger, debt and the many other gross inequities which insult our common humanity.

Jon Tinker
President
The Panos Institute

Introduction

How significant is global warming? Is it a remote and implausible threat which has been blown out of proportion by irresponsible media coverage? Or is it the greatest danger facing the human race?

The purpose of this book is to provide a clear and concise account of the facts and arguments about global warming. It aims to provide the baseline knowledge required to understand the present position and follow the debate on global warming as it evolves in the coming years.

Chapter 1 outlines the facts on global warming as they are known today. Chapter 2 looks at what scientists are predicting for the next century. Chapter 3 looks at the implications, especially for the Third World, if these predictions are realised.

Chapter 4 outlines the actions which can be taken worldwide to reduce the risk of global warming, and Chapter 5 discusses the policy implications for the Third World. The Appendix gives a brief description of the principal international organisations working on global warming and the main actions taken to date.

The author

Gerald Foley has been writing about energy and related issues since 1970. He has travelled widely in the developing world and worked as a consultant to a variety of governments and international organisations.

Acknowledgements

Many thanks are due to Mick Kelly and Sarah Granich for information provided for this book and for invaluable comments on an earlier draft. Others who have helped include Stewart Boyle, Olivia Bennett, James Deane, Daniel Nelson, R K Pachauri, Vandana Shiva and Jon Tinker. Responsibility for the opinions expressed here rests entirely with the author.

ABBREVIATIONS
CFCs: chlorofluorocarbons
GCM: General Circulation Model (of the atmosphere)
ICSU: International Council of Scientific Unions
IPCC: Intergovernmental Panel on Climate Change
NASA: (US) National Aeronautics and Space Administration
ppb: parts per billion by volume
ppm: parts per million by volume
UNEP: United Nations Environment Programme
UNCED: United Nations Conference on Environment and Development
WCP: World Climate Programme
WMO: World Meteorological Organization
Note: "billion" is used to mean one thousand million.

CHAPTER ONE

Global Warming: The Facts So Far

Global warming is not just a question of obscure scientific theory. If it is really taking place on a significant scale, it is of enormous practical importance for the not very distant future of the whole human race.

It is therefore important that the evidence is set out as clearly and soberly as possible. This chapter attempts to summarise those aspects of global warming on which there is now a broad scientific consensus.

What is the greenhouse effect?

There is no doubt about the reality of the greenhouse effect. It is what has made the Earth a home for life and Venus completely uninhabitable.

The greenhouse effect is so-called because of its similarity to what happens in a greenhouse when the sun is shining. Sunshine coming in through the glass roof and walls of the greenhouse heats up the interior so that its temperature becomes higher than that on the outside. The reason is that the glass stops a certain proportion of the heat from escaping. The greenhouse, in other words, acts as a heat trap.

In the case of the Earth, the greenhouse effect is a result of the presence of small quantities of certain gases in the atmosphere. These are called greenhouse gases. Water droplets and ice crystals in clouds and other small particles in the atmosphere also trap heat which would otherwise escape. The gases in the Earth's atmosphere, therefore, act in a similar way to the glass in a greenhouse.

The role of the greenhouse gases has been under discussion by scientists for over 150 years. The first to mention the greenhouse effect was the French scientist Jean-Baptiste Fourier in a paper written in 1827. The Irish-born scientist John Tyndall, in a lecture in London in 1861, showed that the water vapour in the air increased its absorption of heat from the sun by 15 times compared with dry air. Svante Arrhenius, a Swedish scientist, in 1896 calculated the effect of doubling the quantity of carbon dioxide in the atmosphere and obtained results remarkably close to present estimates.

The two nearest planets in the solar system, Mars and Venus, provide graphic illustrations of the greenhouse effect. On Mars, where the greenhouse gases have almost totally disappeared from the atmosphere, the temperature is about -60°C, far too cold for any known form of life. Venus, on the other hand, has an extremely high concentration of greenhouse gases and the temperature is around 480°C, hot enough to melt tin.

The Earth, with its average temperature of about 15°C, falls between these two extremes. Because of the naturally occurring greenhouse gases in the atmosphere, this average temperature is about 33°C higher than it would be without those gases.

This natural greenhouse effect provides the earth with a climate in which plant, animal and human life has been able to thrive. During the last 200 years, however, the quantity of greenhouse gases in the atmosphere has been steadily increasing as a result of human activities.

The action of the greenhouse gases in increasing the temperature is sometimes described as "radiative forcing". It is important to distinguish between the greenhouse effect—a natural process—and the threat posed by the greenhouse gas emissions resulting from human activity. The increase in the greenhouse effect caused by these additional gases is sometimes described as the anthropogenic (generated by humans) or "enhanced" greenhouse effect.

The greenhouse gases

The Earth's atmosphere consists mainly of nitrogen which makes up 78% of the total. Oxygen is the next most common gas and accounts for about 20%. The greenhouse effect of each is slight.

The most important greenhouse gases—those which trap heat inside the atmosphere—are water vapour and carbon dioxide. Others which also occur naturally are methane, nitrous oxide and ozone. In addition, there are several manufactured gases, of which the most important are the chlorofluorocarbons (CFCs), which have an extremely potent greenhouse effect.

Fossil fuel consumption is at the heart of global warming. Concentrations of greenhouse gases have increased dramatically since the beginning of industrialisation in Europe.

Carbon dioxide

Carbon dioxide (CO_2) is the most important of the greenhouse gases being added to the atmosphere by human activity. Its natural concentration is tiny, only about 0.03%, but in addition to its greenhouse effect it also plays a vital natural role in the biosphere.

Carbon dioxide is absorbed by plants and, with the aid of sunlight, broken down to form their tissue in the process known as photosynthesis. A similar process happens in the seas where the carbon dioxide is absorbed by algae. In both cases, oxygen is emitted as a by-product.

During the night, when there is no sunlight, the reverse happens, with plants and algae absorbing oxygen and emitting carbon dioxide. This is known as respiration, or breathing; humans and animals also take in oxygen and emit carbon dioxide when they breathe.

When plants or animals die and decay, the carbon they contain is released in the form of carbon dioxide. Burning wood or fossil fuels similarly releases carbon dioxide. Natural soil also contains carbon; up to 50% of its dry weight may be in the form of partly rotted organic material. When this is turned over by the plough, a certain amount of the carbon escapes to the atmosphere in the form of carbon dioxide.

Although direct measurements of the amount of carbon dioxide in the atmosphere were made in the last century, the first comprehensive global readings only began during the International Geophysical Year in 1957. One series of measurements has been carried out since then at the Mauna Loa Observatory on a mountain top in Hawaii.

The Muana Loa observations give a detailed record stretching back over the last 33 years. They show that during each year the concentration of carbon dioxide in the atmosphere drops in the northern spring because of the absorption by growing plants, and rises in the winter as they decay. But there is also a clear upward trend. The average concentration in 1957 was 315 parts per million by volume (ppm) and by 1988 it had risen to 350 ppm, a rate of increase of about 0.3% per year.

There is also a much longer record of the amount of carbon

Figure 1.1 Concentrations of carbon dioxide and methane

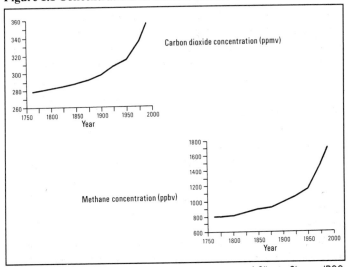

Source: *Policymakers Summary of The Scientific Assessment of Climate Change,* IPCC
Working Group I Report, June 1990
Concentrations of carbon dioxide and methane remained relatively constant up to the eighteenth century, since when they have risen sharply.

dioxide in the air. This comes from a 2.2 km deep borehole drilled at Vostok in the Antarctic. Drilling began in 1980 and the work was completed in 1985. The extracted core of ice was analysed by a team of Soviet and French scientists.

In effect, this was a journey backward in time. The ice at the bottom of the borehole was formed from snow which fell 160,000 years ago. Each subsequent year's snow has formed a distinct band, like the rings in a tree. By analysing the tiny bubbles of air trapped in these bands, it was possible to find out how much carbon dioxide was in the atmosphere at any time over the whole of the period.

The Vostok analysis showed that the present carbon dioxide concentration is higher than any other time during the past 160,000 years. The previous peak of 300 ppm was 135,000 years ago and over most of the period the level was about 280 ppm. Figure 1.1 illustrates how the concentration has changed since the beginning of large-scale industrialisation in 1750. The carbon dioxide concentration is now about 25% higher than in pre-industrial times.

Figure 1.2 The Earth's carbon cycle

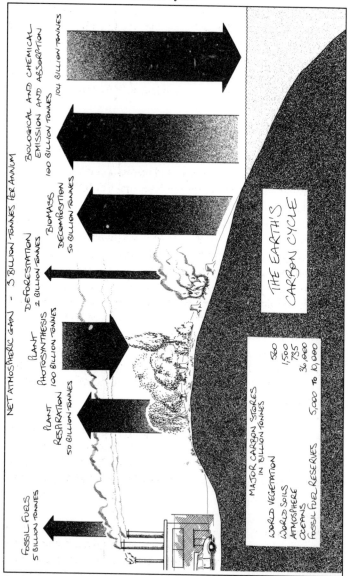

Source: Based on "Global Climate Change", R A Houghton and G M Woodwell, *Scientific American*, April 1989

The carbon dioxide in the atmosphere cannot, however, be considered on its own. It is part of a vast process known as the carbon cycle in which the earth's carbon is continually being shifted between various stores or "sinks". This is illustrated in Figure 1.2. In discussions of the carbon cycle, the figures used normally refer to the amounts of carbon, rather than carbon dioxide. Each tonne of carbon in the atmosphere is equivalent to 3.67 tonnes of carbon dioxide.

It is estimated that there are about 735 billion tonnes of carbon in the atmosphere. Under natural conditions, about 100 billion tonnes are absorbed by plants every year and this is roughly balanced by the amount of decay and respiration taking place.

A similar inflow and outflow of around 100 billion tonnes of carbon takes place in the case of the oceans. Around half the uptake is estimated to be by living creatures with the remainder being absorbed by chemical or physical processes in the top water layers.

The major human contribution to the amount of carbon dioxide in the atmosphere comes from the burning of fossil fuels: coal, petroleum and natural gas. This adds about 5 billion tonnes of carbon to the atmosphere every year. Deforestation and the spread of farming also increase the amount of carbon dioxide in the atmosphere. Although exact calculations are impossible, it is estimated that these two activities are adding 1-2 billion tonnes of carbon to the atmosphere per year. In total, therefore, human activities are adding roughly 6-7 billion tonnes of carbon to the atmosphere every year.

Only about half of this additional carbon stays in the atmosphere, with the remainder being taken up by the oceans and land vegetation. The exact figures are still a matter of debate, with some scientists saying that as much as 50% is taken up by land vegetation. There is also a scientific debate about where the carbon goes, once it is absorbed by land vegetation. Some scientists believe the grasslands can store as much carbon as forests.

In comparison with the quantities of carbon circulating between the atmosphere, the oceans and the biosphere, the amounts stored for longer periods are much larger. The total quantity of carbon in the soils, boglands and tundra of the earth is about 1,500 billion tonnes, over twice as much as that in the atmosphere. A further

Michael Harvey/Panos Pictures

Deforestation in the Amazon. When forests are cleared and burned, not only is carbon dioxide released into the atmosphere, but the world loses valuable carbon "sinks" or stores.

5,000-10,000 billion tonnes of carbon are contained in fossil fuels, mainly coal. The amount in the deep oceans is immense, around 36,000 billion tonnes.

There are still important questions about the carbon cycle which remain unanswered. Many of the details of how carbon is circulated between the atmosphere, the oceans, the soils and land vegetation are not yet clear. Nor can anyone predict with confidence how the whole system may change as the concentration of carbon dioxide in the atmosphere continues to increase.

Methane

Methane (CH_4) is another naturally occurring greenhouse gas. It is produced when certain types of bacteria break down organic material in the absence of air. Methane burns easily, yielding carbon dioxide as a by-product.

Methane is naturally produced during the decay of organic material in swamps and is sometimes called marsh gas. It is also a by-product of the activity of termites; and it is produced in the stomachs of cows and other animals, emerging when they belch and break wind.

Human activity has considerably increased the amount of methane being released into the atmosphere. Paddy fields provide ideal conditions for its formation, with the rice stalks apparently acting as straws to channel the methane into the atmosphere. The growing number of cattle, buffaloes and similar animals is another significant source. Methane is also produced in substantial quantities in rubbish dumps; in some places, the quantity is so great that it is profitable to collect it as fuel for boilers or electric power generation.

Methane is the principal component of natural gas. Huge quantities are found in association with oil or in natural gas fields. It is also found in conjunction with coal and is a deadly hazard in mines; the majority of explosions in coal mines are the result of a mixture of air and methane being ignited by an accidental spark. Small quantities are produced in the burning of wood and other fuels, especially when the combustion is not particularly efficient; significant amounts are also emitted in charcoal-making.

A limited number of measurements have been taken of the amounts of methane generated in paddy fields; only one study has, however, been made in Asia where 85% of the world's rice is grown. From these studies, it is calculated that the annual emissions from rice-growing are about 20% of the total added to the atmosphere each year. Emissions from cattle account for roughly another 20%.

Large quantities of methane are locked up in the frozen tundra in the Arctic. There are also considerable quantities in the mud at the bottom of some oceans. One of the effects of a rise in global temperatures could be to release methane from these sources.

Ice cores from the Greenland ice-cap show that the quantity of methane in the atmosphere was roughly constant at about 0.7 ppm from the end of the last ice age some 10,000 years ago up to about the year 1700. It then began to grow more or less in line with the human population. The reason is not hard to find. As the human population increases, so also do the numbers of cattle, the area of land under rice cultivation and the amount of fuel being burned. In the present century, additional contributions are being made by leakage of methane from natural gas wells, gas distribution systems, oil wells and coal mines.

The concentration of methane in the atmosphere is now about 1.7 ppm. This is almost 2.5 times greater than it was 300 years ago. In recent decades it has been growing at an average rate of just under 1% per year. Methane breaks down relatively easily and is estimated to have an average life of about 10 years in the atmosphere.

Nitrous oxide

Nitrous oxide (N_2O) is another naturally occurring greenhouse gas. It was formerly used widely as a light anaesthetic, sometimes causing people to laugh, and becoming known as "laughing gas".

Not much is known in detail about the origins of the nitrous oxide in the atmosphere. It is believed that the main source, accounting for perhaps 90% of the total, is microbial action in the soil. The application of nitrogenous fertiliser increases the rate at which it is released. It is also produced in small quantities by the burning of fossil fuels. The amount of nitrous oxide in the atmosphere is tiny and is measured in parts per billion (ppb). The concentration in 1986 was about 310 ppb and it was growing at about 0.25% per year.

Figure 1.3 Concentrations of nitrous oxide and CFCs

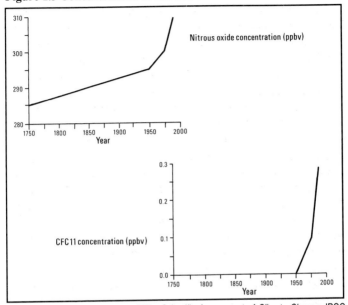

Source: *Policymakers Summary of The Scientific Assessment of Climate Change,* IPCC Working Group I Report, June 1990

Concentrations of nitrous oxide have increased dramatically in the last few decades. CFCs were not present in the atmosphere before the 1930s.

One of the characteristics of nitrous oxide is that it is extremely long lived. A molecule is estimated to last an average of 150 years in the atmosphere. Even small emissions therefore build up steadily.

The CFCs

The chlorofluorocarbons (CFCs) are a family of manufactured gases which were introduced by General Motors, a car manufacturing company, in the United States in the 1930s. The irony is that CFCs possessed so many admirable qualities—they are non-poisonous, non-inflammable and highly stable—that they rapidly found a variety of applications and came into widespread use after the Second World War. The most widely used CFC is made by the chemical company Dupont under the trade name of Freon. They belong to a wider group of gases known as the halocarbons.

Two types of CFCs are in common use: CFC-11 and CFC-12. CFC-11 is principally used in blowing plastic foams and as the propellant in spray cans. CFC-12 is used in air conditioning and refrigeration units as well as in spray cans and for foam-blowing. CFCs are also used as solvents for cleaning computer circuits during manufacture.

The fact that CFCs are so stable means that they are highly persistent: CFC-11 lasts on average 65 years in the atmosphere and CFC-12 about 130 years. The concentration of CFC-11 in the atmosphere is just 0.3 ppb (roughly one millionth that of carbon dioxide) and that of CFC-12 is just under 0.5 ppb—but they are both extremely potent greenhouse gases. Their present rate of increase is about 4% per year.

Ozone

Ozone is a naturally occurring form of oxygen. It has the chemical formula O_3, which means that each molecule consists of three atoms of oxygen rather than the two atoms of ordinary oxygen which has the chemical formula O_2.

Ozone occurs naturally in the stratosphere, the band of the atmosphere stretching from around 10 km to 50 km above the surface of the earth. The ozone in the stratosphere is formed by the interaction of ultraviolet radiation from the sun with the oxygen present in the stratosphere. There is a natural balance between the rate at which the ozone is formed and the rate at which it breaks down so that the quantity in the stratosphere stays roughly constant.

The ozone in the stratosphere, referred to as the ozone layer, absorbs ultraviolet radiation. This is extremely important since ultraviolet radiation is damaging to animal and plant life. In humans it causes localised skin cancers, most of which can be cured provided they are treated; but it also causes melanoma, a far more serious skin cancer, which begins as a mole but rapidly spreads cancer cells throughout the body. Ultraviolet radiation also damages the eyes, and can cause cataracts. It has been suggested that increased exposure to ultraviolet radiation can weaken the immune system. Ultraviolet radiation can reduce the yield of crops and is also damaging to marine algae. In its role of protecting the surface of the earth from ultraviolet radiation, ozone is entirely benign.

At lower levels, in contrast, ozone is very definitely a pollutant. It is one of the by-products formed when sunlight interacts with vehicle exhaust emissions. In heavy smog it reaches levels which are damaging to plant and animal life. It is also produced in industry, where it is used as a bleaching and purifying agent.

Ozone is highly effective as a greenhouse gas. It is, however, relatively unstable, lasting only a few weeks in the atmosphere. Its contribution to the greenhouse effect varies from place to place and the overall impact is difficult to estimate exactly. Nevertheless, it does appear to have a significant and steadily growing effect.

Water vapour, other greenhouse gases, and clouds
A variety of other gases also contribute to the earth's greenhouse effect. They include water vapour and a number of other manufactured gases, including substitutes for CFCs.

Water vapour is invisible and should be distinguished from clouds and mist which are formed when water vapour condenses to form water droplets. Water vapour, in fact, accounts for by far the greater part of the greenhouse effect. The amount of water vapour in the atmosphere is outside direct human control and is largely governed by the global temperature. But if human actions raise that temperature, then there will be an increase in the water vapour.

Water vapour is, therefore, an important factor in determining the final effect of any increase in the greenhouse gases caused by human activities. If the Earth becomes warmer, the quantity of water vapour in the atmosphere will increase because of higher evaporation rates. This will increase the greenhouse effect and give added impetus to any global warming taking place.

Other gases with a greenhouse effect include carbon tetrachloride, which is used as a dry cleaning fluid, and halocarbon gases, some of which are proposed as substitutes for the CFCs. Although the effect of these is still small, it is important that the amounts being added to the atmosphere are kept under strict review.

The role of clouds in the earth's climate system is both complex and extremely important. They reflect a certain proportion of the incoming solar radiation, thus reducing the energy reaching the surface of the earth. They also absorb both incoming and outgoing radiation, a certain proportion of which is radiated back down to the

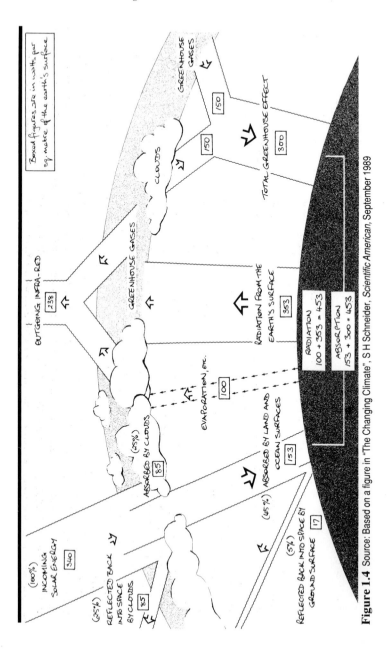

Figure 1.4 Source: Based on a figure in "The Changing Climate", S H Schneider, *Scientific American*, September 1989

The Earth's Energy Balance

The greenhouse effect is part of the complex series of interactions which determine the Earth's energy balance and temperature.

About 30% of the incoming energy from the sun is directly reflected back into space by clouds (25%) or reflective ground surfaces (5%) and plays no further part in the energy balance. About 25% is absorbed by clouds and the remaining 45% is absorbed by land and ocean surfaces which are warmed as a result.

The heat emitted from the surface of the Earth is in the form of invisible infra-red long-wave radiation. This is absorbed by greenhouse gases, clouds and particles in the atmosphere, thereby raising their temperature.

These, in turn, radiate infra-red energy back down to the Earth's surface. Heat is also carried up into the atmosphere by evaporation: rising air and water vapour, and other atmospheric processes.

As shown in the diagram, the effects of clouds and greenhouse gases are approximately equal (150 watts per square metre) with each contributing roughly the same amount as the direct solar radiation (153 watts per square metre). In fact, the combined effect of greenhouse gases and clouds accounts for about two-thirds of the radiation reaching the surface of the Earth.

Looked at another way, if one were to imagine a square metre column of air stretching from the Earth's surface to the top of the atmosphere, the combined effect of the naturally occurring greenhouse gases and the clouds is to add about 300 watts to the energy reaching the Earth's surface. The effect of the 50% increase in the carbon dioxide equivalent which has taken place since pre-industrial times is about 2 watts per square metre.

This static picture of the Earth's energy balance is, however, a major simplification. If the greenhouse effect alone were at work, the earth's temperature would be about 77°C, making the Earth not exactly a Venus, but far too hot for most life. The reason the temperature is at 15°C is because of the action of the atmospheric processes and air circulation which distribute heat around the globe and into the upper atmosphere.

Added to atmospheric redistribution is that of the oceans. Vast amounts of energy absorbed in the equatorial and tropical regions are carried north and south and replaced by cold waters flowing deep along the ocean beds from the polar regions.

surface of the earth. The overall effect of radiation from clouds is broadly equivalent to that of the natural greenhouse effect.

The combined effect

Calculating the combined effect of the various greenhouse gases is a fairly complex process. The impact of each gas depends not just on its greenhouse properties but on its lifetime in the atmosphere.

The greenhouse effect of a certain amount of methane, for example, which is added to the atmosphere today will decline relatively rapidly as the gas is broken down by natural processes. But a similar amount of CFC-12 will still be active in the early part of the twenty-second century. The overall impact of a greenhouse gas, taking into account its radiative effect and its lifetime, is referred to as its global warming potential.

The present concentrations and rates of increase of the principal greenhouse gases in the atmosphere are shown in Table 1.1. The Table also shows the lifetime in the atmosphere and the global warming potential of each gas in comparison with carbon dioxide. The comparison is based upon the effect of 1 kg of each gas over a period of 20 years.

The combined effect of the greenhouse gases so far added to the atmosphere as a result of human activities is equivalent to a 50% rise in the carbon dioxide concentration. Of that, about half is attributable to carbon dioxide itself and the remainder to the other greenhouse gases.

CFCs and the ozone hole

In addition to their role as greenhouse gases, CFCs are also important because they damage the ozone layer of the upper atmosphere. In doing so, they expose humans and other living creatures to the dangers of increased ultraviolet radiation.

In the late 1960s and early 1970s, scientists began to worry about the damage that exhaust emissions from supersonic aircraft (such as Concorde), which fly in the stratosphere, might cause to the ozone layer. As it turned out, these fears were unjustified, but scientists became aware that some of the other substances being released into the atmosphere could be a serious danger.

Table 1.1 Comparison of greenhouse gases

Greenhouse gas	Approximate greenhouse impact compared with carbon dioxide*	Present concentration	Present rate of increase (%/year)	Lifetime in atmosphere (years)
Carbon dioxide	1	353 ppm	0.5	50-200
Methane	60	1.7 ppm	0.9	10
Nitrous oxide	270	0.3 ppm	0.2	150
Ozone	2000	100 ppb	1	0.1
CFC-11	4500	0.3 ppb	4	65
CFC-12	7000	0.5 ppb	4	130
HCFC-22**	4100	n.a.	n.a.	15

* Relative contribution measured over a 20-year period per kilogramme of gas.
** One of the potential substitutes for CFCs

Source: Based on data from IPCC Working Group Reports

The CFCs were among the prime suspects and scientific investigation showed that they were indeed a significant threat, but there was no direct evidence to show that they were actually causing any damage. After much acrimonious debate, however, the use of CFCs in spray cans, perhaps their least essential use, was heavily restricted by the US government in 1978 and other countries gradually followed suit.

In the early 1980s, dramatic proof of damage to the ozone layer was discovered. Members of the British Antarctic Survey, who were taking measurements from the ground, began to notice a decline in the quantity of ozone in the ozone layer. By 1984, when there was a 30% loss, they were certain something serious was happening and their paper announcing this was published in 1985.

Measurements in the following years showed that the depletion was continuing, with a virtual "hole" forming over the South Pole. The existence of the hole was quickly confirmed by analysis of satellite data. By 1987, the hole had spread to cover the whole Antarctic continent. There is now clear evidence that the whole ozone layer has been significantly depleted and that this process is continuing.

The overall reduction reported in mid-1991 was about 8%. The reduction is somewhat less over the equator. Already the world is suffering a significantly increased dosage of ultraviolet radiation.

The discovery of the ozone hole launched a massive international scientific effort to discover what exactly was happening. It was found that CFCs were almost certainly to blame. Even though they were present in extremely small concentrations, they were instrumental in breaking down the molecules of ozone; each molecule of CFC is, in fact, capable of bringing about the destruction of 100,000 molecules of ozone.

In the meantime, efforts to achieve an international agreement on limiting the production of CFCs had been under way. The discovery of the ozone hole accelerated the process and an international treaty to restrict the use of CFCs, the Montreal Protocol on Substances which Deplete the Ozone Layer, was signed in 1988. A further meeting in London in 1990 agreed to speed up the process and phase out CFCs by the year 2000.

One of the emerging problems, however, is that some of the substitutes for CFCs are themselves potent greenhouse gases. The gas HCFC-12, for example, has a greenhouse effect just slightly less than that of CFC-11.

CHAPTER TWO

Looking to the Future

The greenhouse effect is a natural process and a scientifically proven fact. As a result of human activity, the concentration of greenhouse gases in the atmosphere is clearly rising. What does this mean for the world?

Forecasts, scenarios and models

A variety of terms are used, sometimes without a great deal of precision, when the future is being discussed. It is useful to clarify what they mean.

A forecast is an attempt to predict what will actually happen. It uses the available information to make a guess, as accurately as possible, about the future. Weather forecasts, for example, endeavour to predict what the weather is going to be a day, a week, or further into the future.

The further ahead a forecast is made, the more difficult it is to get right. Long-term forecasts of energy consumption, for example, have a dismal record. In the early 1970s, world energy consumption was widely forecast to treble by the 1990s. In the event, it has increased by about 60%.

A scenario, in contrast, does not attempt to describe what the future will be. Its purpose is to analyse the implications if certain things happen or a particular course of action is taken. Thus, a high greenhouse gas scenario can be used to examine the consequences of a major increase in greenhouse gases but without implying that this is necessarily going to happen. Similarly, scenarios in which the sea level rises a couple of metres, or other changes take place, can be used to explore the consequences of those possible changes.

A computer model is neither a forecast nor a scenario. It is a highly simplified version of reality expressed in mathematical terms which can be handled by a computer. It enables scientists to alter some of these mathematical expressions (for example, those representing the greenhouse gases in the atmosphere), feed them into the computer and see what happens.

The mystique of computers is still such that people sometimes credit them with powers of insight and prediction that they do not possess. The credibility of the results of climatic analyses carried out by computer models depends entirely on how well these models mimic the reality of the earth's climatic systems.

General Circulation Models

At the heart of the discussion on global warming lie a number of large and sophisticated computer models called General Circulation Models (GCMs). The amount of data involved is huge and only the largest super-computers are able to handle these models.

There are about half a dozen centres with GCMs at present. Among these are the National Center for Atmospheric Research in Colorado, the Goddard Institute for Space Studies (part of NASA) in New York, the Geophysical Fluid Dynamics Laboratory at Princeton University, the Soviet Hydro-Meteorology Centre in Moscow, and the UK Meteorological Office.

The GCMs divide the earth and the atmosphere into a three-dimensional grid and are able to calculate the way in which temperature, wind, humidity and other variables are likely to change under different conditions. When studying global warming, the model is first run without any increase in greenhouse gases. The concentration of greenhouse gases is then increased and the model is run again. The difference between the two runs is the calculated increase in the greenhouse effect (over the natural level).

One of the major constraints on these model studies is cost. A typical computer run covering a few decades of simulated changes in world climate can take 100 hours. As the cost of running a super-computer may be US$1,000 per hour, there are clearly limits on the amount of investigation which any particular centre can afford.

Ron Giling/Panos Pictures

Drought in Mali. Human activity has undoubtedly contributed to environmental degradation in the Sahel, but some scientists believe the deteriorating conditions may also be attributable to global warming.

Another limitation is that the degree of detail represented in the models, despite their size, is still relatively small. The grid squares on the model used at the National Center for Atmospheric Research, for instance, cover an area of over 300,000 square kilometres each, far larger than many countries, and over twice the size of Bangladesh, for example. The model is only able to deliver a single average figure for important climatic features such as temperature, rainfall and cloudiness over the whole grid. Therefore, the results emerging from such models can tell us little about variations in climate patterns at a national level; the best they can do is reveal broad regional averages.

Climatologists have made major efforts to ensure that the GCMs now in use represent reality as closely as possible. One of the ways of checking on this is to feed in climatic data from the past to see whether they are able to "predict" what actually happened. Thus, they should be able to "predict" the seasons; they should also be able to replicate long-term events such as the emergence from the last ice age. The best GCMs have by now earned the right to be taken seriously.

In addition to the GCMs, climatologists use a wide range of other computer models to provide checks on the results of the GCMs and to investigate different aspects of the climatic system in more detail. There are, for example, simplified global models which calculate only a single average value for the earth's temperature; models which look at the way the oceans transfer heat from one part of the earth to another; models of the carbon cycle; and a series of models which can look in more detail at what is likely to happen at a regional level if the overall predictions of the GCMs are realised.

The Intergovernmental Panel on Climate Change

The best known and most prestigious set of initials in the global warming discussion is that belonging to the Intergovernmental Panel on Climate Change (IPCC). This was set up under a joint initiative by the World Meteorological Organization and UNEP in 1988. At its first session, in November 1988, the IPCC set up three working groups to examine the global warming issue.

Working Group I was responsible for assessing the scientific information on greenhouse gas emissions, global warming, the climate models being used, and the projections of global warming and climate change. Working Group II reviewed the environmental and socio-economic impacts of climate change. Working Group III analysed possible policy responses and actions which might be taken to reduce the threat of climate change or deal with its effects if it took place.

More than 1,000 scientists from all over the world have been involved in one way or another with the work of the IPCC. Its second working session took place in Nairobi (Kenya) in June 1989 and the third in Washington DC (USA) in February 1990. Summaries of the reports of the three Working Groups became available in June and July of 1990. They represent the best expert consensus to date on the global warming issue.

The Working Groups' reports have provided the impetus for negotiations on a climate treaty which it is hoped will be completed and signed at the 1992 United Nations Conference on Environment and Development (UNCED) in Brazil (see p70). A key sticking point in the run-up to this summit, and in the climate negotiations themselves, is how to reduce emissions without slowing down the drive for a higher standard of living in developing countries.

"Business-as-usual"

The human race has a variety of choices as far as the build-up of greenhouse gases in the atmosphere is concerned. One is to take little or no action to alter the present trends of steadily rising emissions of carbon dioxide and other gases. This is often referred to as the "business-as-usual" scenario.

Rising carbon dioxide emissions

The present world consumption of commercial energy is about 8 billion tonnes of oil equivalent per year. Of this, just under 40% is oil, 27% is coal, and 22% is natural gas, with hydro-electric and nuclear power making up the remaining 11%. There are no obvious physical or resource limitations which are likely to restrict consumption of fossil fuels over the next few decades.

There is enough oil in already proven reserves for another 45 years production at present rates. This is assuming that not another drop is discovered—and everyone agrees there is plenty of oil waiting to be found. Nor, short of another and more widespread Middle East war, does there appear to be any serious prospect of prices rising greatly in real terms over the next decade or so.

The supply position for natural gas is equally assured. Proven reserves are sufficient for 60 years production at present levels. Experts are unanimous that there are vast quantities yet to be discovered. The main constraint on expansion of natural gas consumption, especially in developing countries, is likely to be a shortage of the investment capital required to build pipelines and distribution systems.

Coal is also abundant. The present estimated economically recoverable reserves are sufficient to sustain today's level of consumption for the next 300 years. The total amount of coal in the ground could probably keep present consumption going for another 2,000 years.

The preponderance of coal in the energy reserve picture is bad news for the environment. In addition to being the most generally

David Reed/Panos Pictures

Worldwide, coal is the dominant fuel and is likely to remain so, especially in China and India. But the emissions resulting from its use stem primarily from the industrial world.

polluting of the fossil fuels, it has the highest output of carbon dioxide per unit of energy. Burning a tonne of coal produces about 2.5 tonnes of carbon dioxide. To obtain the same amount of energy from oil, the amount of carbon dioxide emitted would be 2 tonnes and from natural gas it would be only 1.5 tonnes. Wood is distinctly worse, emitting about 3.4 tonnes of carbon dioxide to provide the same quantity of energy as a tonne of coal.

Neither hydro nor nuclear power directly emit carbon dioxide when producing electricity. Both, however, are relatively minor energy sources on a global scale; hydro supplies about 6% of the world total and nuclear about 5%. There are major constraints on the expansion of each and neither is likely to increase its share significantly over the next couple of decades.

The contribution of renewable energy sources such as wind and solar is at present negligible on a global scale. Even if there is a major effort to increase the use of these energy sources over the next couple of decades, the dominance of the fossil fuels is likely to continue until at least the middle of the next century.

Over the past 20 years, world energy consumption has grown at an average rate of about 2.4% per year, but during the 1980s, the growth rate fell to about 1.8% per year. The choice as to what exactly constitutes business-as-usual is thus fairly open. But simply projecting the 1980s growth rate forward would bring a doubling of energy consumption by about the year 2030.

Present carbon dioxide emissions from deforestation and farming are estimated to be in the range of 1-2 billion tonnes per year. Under a business-as-usual scenario, these emissions are likely to rise, contributing proportionally to the predicted doubling of fossil fuel emissions by 2030.

But unlike fossil fuel consumption, there are severe resource constraints emerging in certain areas. Deforestation has progressed so far in some places that it cannot continue at the same rate in the future. Similarly, the expansion of farming is running up against land shortages. Carbon dioxide contributions from these sources may therefore begin to level off in the early decades of the next century. If all the tropical forests were removed, it is estimated that the increase in atmospheric carbon dioxide would be 35-60 ppm, roughly 10-20% over today's levels.

Trends in other greenhouse gas emissions

Data on present emissions of nitrous oxide and methane are extremely poor. As a result, there is no basis for accurate predictions of their use in the future. Emission levels, nevertheless, appear to be linked to the number of people, so emissions are likely to continue rising broadly in line with population growth.

In the case of CFCs, since an international agreement to end their use by the year 2000 has been signed, the future level of emissions will depend primarily upon how strictly the agreement is observed. But the future concentration also depends on what happens to the substantial amounts of CFCs contained in the cooling circuits of the vast numbers of refrigerators and air conditioning units already in existence. Rigorous measures will be needed to prevent these CFCs being released to the atmosphere when the equipment is scrapped.

The IPCC carried out a number of analyses of the likely future emissions of nitrous oxide, methane and CFCs. Its conclusion was that the combined effect of these three types of gas was likely to equal that of carbon dioxide emissions over the next 30-40 years.

Doubled greenhouse gas concentration by 2030

Putting the projected contributions of carbon dioxide and the other gases together, the IPCC's business-as-usual scenario envisages an increase in greenhouse gases equivalent to a doubling of the carbon dioxide content of the atmosphere from pre-industrial levels by about the year 2030. Of that total, around half is projected to be in the form of carbon dioxide itself, with the combined effect of the other gases making up the other half.

The potential for error in the details of any such projection is evident, but the broad conclusion is reasonably robust. With increases in all the main greenhouse gases highly likely unless major efforts are made to curtail them, the only question is when exactly the doubling occurs. A higher level of emissions than that projected could bring the doubling forward to the year 2020; a lower level of emissions could delay it until 2040 or 2050 or possibly longer. But there is no doubt that the world is firmly on course for a doubling of the equivalent carbon dioxide content of the atmosphere sometime in the early to middle decades of the next century.

Table 2.1 Estimates of relative contributions to the greenhouse effect by sector and gas, 1980-2030

Sector	Carbon dioxide	Methane	Ozone	Nitrous oxide	CFCs	Sectoral contribution (%)
Energy	35	4	6	4	0	49
Deforestation	10	4	0	0	0	14
Agriculture	3	8	0	2	0	13
Industry	2	0	2	0	20	24
% by each gas	50	16	8	6	20	100

Source: UNEP/Beijer Institute, 1989

Table 2.1 gives a rough breakdown of the likely contribution of each gas to the greenhouse effect over the period 1980-2030. The table also shows the contributions from different sectors of human activity.

Energy consumption is the dominant sector of activity and is also responsible for roughly half the effect. The projection to the year 2030 does not, however, represent the end of the process. Even if the growth in emissions were to stop completely at that stage, the rate at which greenhouse gases were being added to the atmosphere would be considerably higher than the rate at which they were being removed; this is also true of today's emissions. To achieve a stable concentration of greenhouse gases, a major reduction of the present, not just the future, rate of emissions would be required.

What the computers predict

Having constructed a business-as-usual scenario for the next half century or so of greenhouse gas emissions, the next question is: what does this mean for the global climate? This is where the GCMs are used.

A variety of model analyses using the GCMs have been carried out. The results point broadly to the same conclusion: a rise in global temperatures. The full extent of this rise would not, however, be felt immediately but would be delayed because of the "thermal inertia"—slowness to heat up—of the oceans. The actual warming anticipated at any given time is described as the "realised" warming. The subsequent warming which takes place over the following decades is referred to as the "equilibrium" or "committed" warming.

The IPCC, summarising its analysis of the available computer modelling results, gave a "best guess" prediction of a realised temperature rise of 1°C over today's by about the year 2030 and a total rise of 3°C towards the end of the century. The range of uncertainty for the year 2030 was put at 0.5-1.5°C and for the end of the century at 1.5-4.5°C.

The models predict that the surface air will warm more quickly over land than over the oceans. The warming is also predicted to be higher than the average in the high northern latitudes and lower in the tropics; it is likely to be greater in winter than in summer.

As a result of higher temperatures and greater evaporation from the oceans, rainfall will tend to increase. The IPCC has said the rise by 2030 would be "a few per cent". This would not be evenly distributed. Some of the models suggest it would occur mainly in the higher latitudes and the wetter tropical areas, with some of the arid and semi-arid areas having a reduction in rainfall. It is also believed that the increased rainfall would tend to fall in coastal areas, leaving the continental interiors hotter and drier.

In addition to examining the GCM results, the IPCC carried out a more detailed analysis of the possible climatic changes by the year 2030 for five areas of the world. These studies used larger scale climatic models and incorporated the predictions from the GCMs. The IPCC emphasised that confidence in the results of these regional studies was low.

The study for the Sahel suggested a warming of 0.2-2.2°C with an increase in average rainfall and a marginal decrease in soil moisture in summer, but the analysis showed that there would be both increases and decreases in rainfall and soil moisture throughout the region. The analysis for southern Asia gave an average warming of 0.2-1.2°C with little change in winter rainfall but a 10-15% increase in summer.

One of the direct consequences of an increase in global temperatures would be a rise in sea levels. There are two basic reasons why this would happen. Firstly, the oceans would expand as it got warmer; and secondly, there would be a gradual melting of glaciers and ice sheets on the land. Both of these processes are subject to considerable uncertainties and there is a wide range in the estimates of the contribution of each.

Within the range of temperature increases adopted by the IPCC, the "best guess" prediction of sea level rise by 2030 is 20 cm, with a range of uncertainty of 5-45 cm. Because of the committed warming, there would be a continued rise, giving a total in the range 30-100 cm by the end of the century with a "best guess" of 65 cm.

Some of the popular discussion of global warming has centred on fears that it may result in sea level changes of many metres as a result of the melting of the Antarctic and Greenland ice caps. Such consequences are only likely on a timescale measured in hundreds of years and with temperature rises considerably greater than those presently being predicted.

It is even possible that in the more humid world which would result from global warming, there would be increased snowfall in the Antarctic which is at present almost a desert, receiving very little rain or snow. This could lead to an initial fall in sea level. The melting of floating ice would, however, have no effect on sea level, any more than melting ice in a glass of water causes it to overflow.

A further consequence of global warming could be greater instability in the climate. Because of the higher temperature there would be a greater amount of energy in the atmosphere. This could make extreme climatic events, such as cyclones, storms, droughts, floods and heatwaves, more likely.

Is global warming already under way?

The increase in greenhouse gases in the atmosphere has been under way for well over a century. An obvious question is whether it has yet had any effect.

One well-known climatologist, James Hansen of NASA's Goddard Institute for Space Research in New York, is not in much doubt. At a meeting of the US Senate's Energy Committee in June 1988, he declared that he was 99% certain that the global warming which happened during the 1980s was not a chance event, adding that "it is time to stop waffling so much and say that the evidence

Marc French/Panos Pictures

The aftermath of Hurricane Gilbert, Jamaica, 1989. The 1980s saw an increase in extreme climatic events. Some believe that this indicates global warming is under way, but a connection has yet to be demonstrated.

is pretty strong that the greenhouse effect is here." However, many of his fellow scientists thought that his statement was not warranted by the scientific evidence.

A less stable climate?

Some commentators have suggested that a variety of events during the past decade or so provide a strong indication that the climate is becoming more unstable.

In 1987, for example, there were record high temperatures in Siberia, Eastern Europe and North America. Records were again broken in the same areas the following year. There were terrible floods in Korea and Bangladesh in 1987. In 1988, Bangladesh was hit by floods once more and there was huge loss of life from a cyclone in early 1991. The Maldives were flooded by tidal waves in 1987. The list keeps lengthening.

Although such occurrences are a likely consequence if global warming were under way, it is impossible to say whether they indicate that it is now happening. Winds, floods, storms and other such events depend on a large number of factors, each of which can vary considerably. Chance alone will ensure that they combine to break new records from time to time. While there is, as yet, no direct evidence to link any of these events with global warming, or indeed to indicate that the climate has become more variable, such a connection cannot be dismissed.

A rising trend in global temperatures

It is not easy to measure the temperature of the earth's atmosphere. In any given spot, the temperature varies between the seasons, between day and night, and from minute to minute.

It is also difficult to identify historical trends because of the many ways in which errors could have distorted readings. Non-standard instruments may have been used, or the way in which measurements were taken may have varied as time passed. Ocean temperature readings, for example, used to be taken by hauling up a bucket of water on the end of a rope; now they are taken using water admitted through a tube below the water level. The development of an urban area around a temperature measuring station can alter the average temperature of the area by a couple of degrees.

Another problem is obtaining sufficient coverage. There is no single representative spot which defines the world's temperature. The only way in which a meaningful global temperature can be measured is by taking the average of a large number of standardised

Figure 2.1 Global temperature variation 1861-1989

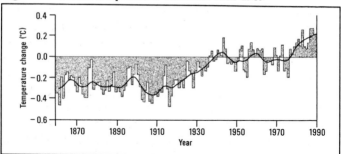

Source: *Policymakers Summary of The Scientific Assessment of Climate Change,* IPCC
Working Group I Report, June 1990

measurements scattered over the whole globe.

Several teams of researchers have collected and examined all the
available temperature readings from different parts of the world.
These records have been analysed and corrected for known sources
of error. The results of the different investigations have been
checked and compared in a number of ways and it is now felt that
they are reasonably close to the truth.

Figure 2.1 shows the world average temperature record over the
period 1861-1989, presented by the IPCC scientific working group.
This is based on records from land stations and ships. The line
wavers up and down; the average temperature dropped sharply
around the turn of the century, rose until the 1940s, varied up and
down and rose sharply in the late 1980s. The overall trend is,
however, clearly upwards. The IPCC verdict is that a real global
warming of 0.3-0.6°C has taken place over the period.

The 1980s have been a period of particularly rapid warming. The
years 1987 and 1988 had the highest global average temperatures
yet measured. Six of the 10 warmest years have occurred during the
1980s. Preliminary data for 1990 show that it topped the previous
record.

But this overall warming has not been experienced uniformly
across the whole earth. During the past 20 years, parts of Europe,
Canada and Greenland, as well as much of Antarctica have, for
example, become cooler. In contrast, parts of southern Asia, North
Africa and the USSR have become warmer. Such regional

variations can have a variety of causes; they can, for example, be a result of changes in the jet stream winds in the stratosphere, and are quite compatible with an overall rise in global temperatures.

Few scientists are yet prepared to attribute the rise in temperature to the onset of global warming. The fact is that the global temperature can vary for a variety of reasons. There was a global warm period from about 900 to 1100 AD during which Viking explorers were able to sail across ice-free northern seas to America, and vines were cultivated several hundred kilometres north of their present range. This was followed by what is called the "little ice age" which reached its coldest in the fourteenth century. Because such natural variations occur, it is impossible to attribute the last century's temperatures rise definitely to the increase in greenhouse gases in the atmosphere.

Researchers have fed the rise in greenhouse gases which has already taken place into a number of GCMs to see how closely they reproduce what has actually happened. The temperature increases calculated by the models were in the range 0.4-1.1°C. The actual warming of about 0.5°C is at the lower end of this range. Whether this represents a tendency to overestimate in the models, an underestimate in the calculations of the rise that has taken place, a counter balancing cooling from volcanic dust, or some other factor, is still impossible to say.

Evidence of rising sea levels

Measuring the level of the seas has many of the same type of problems as measuring the temperature of the atmosphere. Tides, winds, and high and low barometric pressures, all affect the readings of sea levels made at gauging stations.

Another distorting factor is the changes in land levels which occurred after the removal of ice from the last ice age, and is still going on in some places. Other areas, like the Andes, are rising because of geological activity. All such changes have to be identified and compensated for in order to arrive at a true record of global sea levels.

A number of studies have analysed the available data on sea levels. There is broad agreement that the levels have risen over the past century with the IPCC estimating a rise in the range 10-20 cm.

Again, these results are compatible with the onset of global warming. But it is impossible to say whether they are a result of the increase in greenhouse gases in the atmosphere which has taken place to date.

Remaining uncertainties

Projections into the future are always uncertain. In the case of global warming, the uncertainties can be grouped under two headings. The first are the scientific uncertainties and are a result of limitations in the data and in the models used in analysis.

The second type of uncertainties arise from the fact that the future is to a certain extent under human control. As in the case of the CFCs, the human race can take a decision not to proceed on a business-as-usual path and agree on measures to restrict the emission of greenhouse gases.

The level of scientific uncertainty in the global warming discussion is still high at this stage. The carbon cycle, for example, is not yet fully balanced; about 10-20% of the carbon is being absorbed in a sink which has not been identified. There is also considerable debate among scientists about the effects of various possible feedback mechanisms which could either accelerate or slow down the pace of global warming.

Climate models are only as good as our understanding of the processes they describe and this is far from perfect. The ranges in the climate predictions reflect the uncertainties due to model imperfections; the largest of these is cloud feedback (those factors affecting the cloud amount and distribution and the interaction of clouds with solar and terrestrial radiation). Other uncertainties arise from the transfer of energy between the atmosphere and land surfaces, and between the upper and deep layers of the ocean. The treatment of sea-ice in the models is also crude. Nevertheless we have substantial confidence that we can predict at least the broadscale features of temperature change but not of other climatic parameters (eg precipitation, soil moisture) and we are confident that the existing reliability of the predictions can be improved by further research and increased computing capacity.

Overview and Conclusions, *Climate Change: A Key Global Issue,* IPCC Report, July 1990

Flooding in Bangladesh, one of the low-lying countries most at risk from a rise in sea level.

The question of clouds is one example of the complexities involved. Increased cloud cover as a result of greater evaporation will reflect incoming solar energy back into space and hence tend to diminish the greenhouse effect; on the other hand, clouds tend to trap outgoing radiation and hence increase the effect. There is also debate about the effect of vertical circulation of water vapour in the atmosphere, with some scientists suggesting that it may greatly reduce any rise in the global temperature.

A particular area of weakness in the models is their treatment of the oceans. These absorb huge quantities of both heat and carbon dioxide and could profoundly influence the course of any global warming. Yet most GCMs can only manage a crude approximation of the role of the oceans. What effect this has on the reliability of their predictions remains unknown.

Also under discussion is the effect on vegetation of increased carbon dioxide and higher temperatures. These conditions may stimulate more rapid growth and a greater absorption of carbon dioxide, thus damping down the greenhouse effect. Alternatively, it is possible that the more rapid rates of decay and respiration may lead to a net reduction in the amounts of carbon stored.

There is, therefore, considerable room for argument about what is likely to happen. But it is important to remember that any compensating effect which occurs as a result of an increase in the greenhouse effect, while it may reduce the amount of global warming, cannot completely negate it. If the world continues on its present course, the crucial questions are not about whether global warming will happen but about the magnitude and timing of the temperature rise and the impact it will have.

Also, uncertainty cuts both ways. While there are factors which might damp down global warming, there are others which could well accentuate it. There are, for example, large stores of carbon dioxide and methane trapped in the Arctic tundra. If the release of these gases were triggered by a rise in global temperatures, their presence in the atmosphere could quickly accelerate the warming.

Nor does the fact that the global carbon cycle is not yet fully understood provide grounds for the assumption that it will continue to mop up half the carbon dioxide currently being emitted. If global temperatures rise, the amounts absorbed by the oceans may well be reduced, giving an additional impetus to the warming process. The IPCC comment on the range of possible feedback mechanisms was: "Although many of these feedback processes are poorly understood, it seems likely that, overall, they will act to increase, rather than decrease, greenhouse gas concentrations in a warmer world."

It is therefore true that there are grounds for questioning the details of the present understanding of the global warming issue. But it is also important to bear in mind that there is little dispute about the basic underlying science. The IPCC predictions, with all their reservations at the level of detail, are based upon extremely broad consensus among the world's climate scientists.

The natural reaction among policymakers to the fact that there are still uncertainties at a detailed level is to wait until they are resolved. This is not necessarily the right response. It depends on the risks and the cost of reducing them. What are the consequences if indeed there is going to be a rise in global temperatures of 1°C in the next 30-40 years rising to 3°C by the turn of the next century? Is it possible to reduce the level of risk? What are the costs of doing so? These questions will be addressed in the rest of this book.

CHAPTER THREE

The Implications of Global Warming

A rise of 1°C in the average global temperature within the next 40 years (the IPCC conclusion) may not sound very much. It would, however, almost certainly have seriously disruptive effects. The Third World would be particularly vulnerable because it is short of the technical and financial resources required for protection against such impacts.

Of even greater concern is that a realised warming of 1°C by the year 2030 would almost inevitably imply a continuing rise to around 3°C by the end of the twenty-first century. There is no precedent in recorded human history for such a temperature increase. If it happens, it will take the temperature higher than it has been for the past 120,000 years.

Another important aspect of the projected temperature rise is its speed. Where temperatures have changed significantly in the past, it has tended to be over periods of hundreds or thousands of years. A rise of 3°C in just over a century is outside all previous experience.

Food security

Given the lack of firm information, the detailed impact of global warming on agriculture is impossible to assess properly at this stage. Only the broadest generalisations can be made.

However, it is clear that, if the climate changes as predicted, the prospects are, at best, mixed and, in many cases, gloomy. Any reduction in rainfall would obviously be disastrous for poor farmers in the arid and semi-arid areas of sub-Saharan Africa, northeast

Brazil, and parts of India and Pakistan. Countries which depend on monsoon rains would be in dire trouble if there were any major shift in the direction of the monsoon winds or the quantity of rain they brought.

As the Commonwealth Group of Experts appointed to assess the global warming issue remarked about agriculture in the semi-arid tropical regions:

> The vulnerability of this agriculture to climate change derives not only from high levels of dependence on a narrow range of rainfall levels and patterns but also on the characteristic conditions of poverty, high population growth, and environmental stress—including erosion and desertification; these conditions make adjustment to different agricultural practices difficult. The semi-arid regions are already characterised by widespread malnourishment and declining per capita food production, and climate change could aggravate these if it were to increase the frequency or severity of drought [1].

The effects would not be confined to the arid areas. Global warming could make the Mid-West of the United States hotter and windier with a likelihood of dust-bowl conditions; a foretaste of what could be in store was provided by the drought and high temperatures of 1988 when grain yields fell by 30%. Such falls in crop production, if sustained, would almost certainly bring large price rises on international grain markets, which would have severe consequences for Third World and other countries which depend on imports of grain from the United States.

There would be a need to abandon certain crops in some areas. Wheat, for example, because of the increasing shift to bread as a basic element in the urban diet, is being grown ever more widely and often at the margin of its range. Production in these areas would become more difficult, and impossible in some cases.

One of the biggest problems could be the rate of change. Farmers everywhere have shown themselves capable of major adaptations in response to changing circumstances. They are prepared to switch crops as markets change, adopt new seed varieties when they see that these bring advantages, alter cultivation techniques, or take

The effects of global warming will bear heavily on farmers in the Third World, unless they are provided with the resources to adapt agricultural production to changing temperature and rainfall patterns.

whatever measures are likely to increase their security or raise their incomes.

But such adaptations need time and money. If the world is heading into a century of steadily rising global temperatures, the pace and continuity of change is going to place heavy burdens on farmers everywhere. Each decade will bring its own requirements for further adaptation and new investment. While these requirements may be met by the rich farmers of the industrial world, they will bear heavily on the smaller and poorer, and are likely to be too much for the marginal subsistence farmers in many parts of the Third World.

The outlook is, however, not entirely negative. It is, for example, probable that conditions in certain areas will become more favourable for crop growing than at present. One candidate for such climatic improvement is the republic of Russia in the USSR. It is estimated that the higher temperatures and increased rainfall likely in the region would increase grain yields by up to 50%. This would enable the USSR to become a major grain exporter, rather than

Sufficient evidence is now available from a variety of different studies to indicate that changes of climate would have an important effect on agriculture and livestock. Studies have not yet conclusively determined whether, on average, global agricultural potential will increase or decrease. Negative impacts could be felt at the regional level as a result of changes in weather and pests associated with climate change, and changes in ground-level ozone associated with pollutants, necessitating innovations in technology and agricultural management practices. There may be severe effects in some regions, particularly a decline in production in regions of high present-day vulnerability that are least able to adjust. These include Brazil, Peru, the Sahel region of Africa, Southeast Asia, the Asian region of the USSR and China.

Policymakers Summary of the Potential Impacts of Climate Change, IPCC Working Group II Report, June 1990

continuing to be dependent on US imports.

Some commentators have also pointed out that increased carbon dioxide in the air makes crops grow better. Under laboratory conditions, some of the main food and cash crops, such as wheat, potatoes, cotton, rice, barley, cassava and many fruits and vegetables, show an increased yield of 10-50% when the carbon dioxide content of the air is doubled. This has been seen as a potential benefit which could offset some of the damaging effects of global warming.

However, the larger and faster-growing plants would require more water and nutrients. If these were not available in the necessary increased quantities, the productivity of the plants could well be lower than today's. Increased carbon dioxide also makes weeds grow better. It would thus heighten the competition between crops and weeds for the available water and nutrients.

It may well be that a hotter, moister and carbon dioxide-rich world would eventually be able to produce more food than it does today. But before that could happen, it is likely that there would be major disruption of present farming patterns in a large number of areas. The immediate and medium-term impact of global warming would almost certainly be a substantial reduction in food security for probably the majority of the world's people.

Sea level changes

Many of the world's richest and most heavily populated agricultural zones are in low-lying lands along the sea coasts. In total, it is estimated that about half the human race lives in such areas.

They include the deltas of great rivers such as the Ganges-Brahmaputra in Bangladesh, the Nile in Egypt, the Mekong in Indo-China, the Indus in Pakistan and the Yangtse and Hwang Ho in China. Other low-lying areas vulnerable to rising sea levels are in Guyana, Papua New Guinea, eastern Africa, India and Indonesia.

The entire land surface of the Maldives in the Indian Ocean, and some Pacific island countries such as Kiribati and Tuvalu, is only a few metres above sea level. Much of the Netherlands is 5 metres below sea level. Many of the world's largest cities, including Calcutta, Shanghai, Bangkok, Jakarta, Tokyo, London, New York, Miami, Venice and New Orleans, are also in low-level coastal areas.

Any rise in sea level increases the risk of flooding. This is particularly the case if global warming is associated with fiercer storms and hurricanes. Some of the worst disasters in Bangladesh, such as that which killed 250,000 people in 1970, and that which claimed a similar number of victims in 1991, were associated with cyclones. When the sea broke the coastal defences in the Netherlands and the east coast of Britain in 1953, it was a result of major storms in the North Sea.

The IPCC is chilling in its projections:

> ...a rise of sea level by 30-50 cm would affect the habitability of low-lying coastal regions significantly and a one metre rise would impact 360,000 km of coastline, render some island countries uninhabitable, displace tens of millions of people, threaten low-lying urban areas, flood productive land and contaminate fresh water supplies.

A 50 cm rise, for example, would cover an area of Egypt containing a sixth of its total population and farming land. A one metre rise in sea level would cover 14% of Bangladesh, displacing 10% of its people and 14% of its agriculture; physical infrastructure destroyed would include 1.9 million homes, 1,500 km of railway, 10,300 bridges, and 700 km of metalled road. The Sunderbans mangrove forests covering over 400,000 hectares would be destroyed by increasing salinity and then inundation. The total economic output lost would be equivalent to 13% of the country's GDP.

The following description of one of the islands along the coast of Bangladesh is taken from a study of the effects of sea level rise on Bangladesh carried out by the Bangladesh Centre for Advanced Studies. It graphically illustrates the mortal dangers facing millions of people along the coast of the country.

> The entire island is situated only a few feet above the high water mark with a large part of it just above the low water mark (ie in the intertidal zone)....The island is inhabited by over 3,000 people and has a market and a school. However, the only government official on the island is the Beat Officer of the Forest Department and the only two-storied construction is a Cyclone Shelter built by the Red Cross. The major cyclone and tidal wave...in 1982...devastated the island and killed nearly all the inhabitants [2].

Guyana has a population of 900,000, of which 90% lives in the coastal plain. This area is generally below high tide level and is already liable to inundation by the sea as well as flooding from the hills behind it. Sea defences were erected in the late eighteenth and early nineteenth centuries, and strengthened later, but are now in a poor state. Without improved coastal protection, rising sea levels would mean a progressive loss of the coastal agricultural land and a collapse in the country's economy and ability to feed itself.

A number of countries such as the Maldives, and the Pacific island states of Kiribati and Tuvalu, face a frightening future. In the past, storms have driven waves up to 8 metres high right across the land surface of some of the islands. During Hurricane Bebe in 1972, waves up to 15 metres high swept on to the island of Tuvalu smashing buildings, infrastructure, plantations and natural forests. Higher sea levels and a more unstable climate would increase both the impact and the frequency of such events.

Many Third World countries rely heavily on the tourist trade. One of their major attractions is that they offer sweeping expanses of clean sandy beach. Already, many of these beaches are eroding; rising sea levels would mean a rapid increase in the rate at which this is happening. As a rough average, a rise in sea level of 10 cm means a loss of about 10 metres of beach. Left to itself, nature may form a new beach further inland as erosion proceeds. But many areas have coastal defence works to prevent this happening and protect the properties and farming lands immediately behind the beach.

Rising sea levels push further inland the boundary between salt and fresh water in river estuaries. A rise of 10 cm will tend to result in salt water penetrating about one kilometre further inland in a flat estuary. This could have major implications for agriculture, fishing and wildlife habitats.

Intrusion of salt water into fresh water reserves can be a serious problem as sea levels rise. This is compounded as economic growth and larger populations are making higher demands on ground-water resources. Damage from salt water intrusion into the ground-water has already been noted in a large number of places. It threatens supplies of water for drinking and irrigation and will become worse if sea levels rise.

Coral reefs protect many coastal areas, but they are already threatened in several ways. Rising sea levels could outstrip the rate at which the reefs are growing, thus reducing the degree of protection they provide to the coast.

Where the resources are available, it is possible to cope reasonably well with a moderate rise in sea level. Sea defences can be used to protect cities and coastal lands, as in the Netherlands, for

---

Finally, we choose not to argue for wholly negative impacts. There is a realistic expectation that certain positive benefits may accrue; the local response to global change is simply not predictable at this time. What may be perceived as negative to one sector of the wider Caribbean Region may be beneficial to another. Two examples: a change in precipitation associated with a temperature rise may allow the introduction of different crops but perhaps at the sacrifice of others; an increase in the along-shore component of the wind could increase upwelling and be a benefit to fisheries, yet it may be a cause for concern to agronomists dealing with aerial erosion.

"Implications of climatic changes in the wider Caribbean Region. A report by the Task Team of Experts", UNEP, Meeting of Experts on the Caribbean Environment Programme, Mexico 1988

example, over hundreds of years. But such measures require time, money and a reasonable degree of certainty about what is happening.

The projected rise of around 6.5 cm per decade over the next century is rapid in relation to the time required for the installation of major coastal defence systems. The design and construction of hundreds of kilometres of dykes, drainage channels, pumping stations and the rest can only be undertaken on a time-scale of decades. The improvements in Dutch sea defences, put in hand after the 1953 disaster, are only scheduled for completion in the 1990s. The costs are also huge, far beyond the ability of most Third World countries to finance themselves.

Ecological disruption

The plant and animal life in any particular area is biologically adapted to the prevailing climate. Tree species such as the acacia and the baobab, for example, are suited to the rainfall and temperature of the savanna regions. If the climate becomes hotter and drier, these species will give way to a sparse low scrub which is better able to deal with the harsher conditions. Similarly, if the rainfall increases, the savanna species will be superseded by others better able to utilise the increased supply of water.

Changing temperatures and rainfall patterns will therefore impose a variety of pressures upon the plant and animal life of the different ecological zones. If the change is slow, there will be a gradual adaptation to the new conditions, as has happened in the past. It is estimated that, other things being equal, vegetation needs to move 100-150 km towards the poles to compensate for a temperature rise of 1°C.

But, once again, the time-scale is crucial. Trees live for hundreds of years. The natural "movement" of forests is only a couple of kilometres per decade whereas the pace of change in conditions envisaged as a result of global warming is many times greater. In many cases, instead of forests gradually shifting under such rising temperatures, they would simply die.

The same would happen in mangrove areas as a result of rising sea levels. Mangroves are sensitive to changes in sea level, as well

Bruce Paton/Panos Pictures

Dying mangroves, Nigeria. Plant and animal life can adapt to environmental changes, but not if the pace of change is too fast.

as to the changes in water salinity and sedimentation rates which are inevitable with rising sea levels. When this happened in the past, there was time for natural adjustment and the development of new mangrove swamps. But in only a few decades, this would be impossible in most areas.

Coral reefs are among the planet's most diverse ecosystems. One reef can support as many as 3,000 species of marine life. Reefs are particularly vulnerable to any changes in their environment. Extremes in water salinity, temperature, sedimentation, light intensity or pollution can cause the sensitive symbiotic algae, which give corals their colour and food, to be expelled. When this happens, the white limestone skeleton of the corals is exposed, giving it a bleached appearance. Corals usually recover their algae after a bleaching incident, but repeated or prolonged incidents will prevent corals from growing or reproducing and eventually will kill them.

In testimony before a US Senate committee in 1990, several scientists presented strong evidence that recent coral bleaching incidents around the world were directly related to abnormally high ocean temperatures caused, they suspected, by global warming. When the bleaching incidents were compared with global

The rotation period of forests is long and current forests will mature and decline during a climate in which they are increasingly more poorly adapted. Actual impacts depend on the physiological adaptability of trees and the host-parasite relationship. Large losses from both factors in the form of forest declines can occur. Losses from wildfire will be increasingly extensive. The climate zones which control species will move poleward and to higher elevations. Managed forests require large inputs in terms of choice of seedlot and spacing, thinning and protection. They provide a variety of products from fuel to food. The degree of dependency on products varies among countries as does the ability to cope with and withstand loss. The most sensitive areas will be where species are close to their biological limits in terms of temperature and moisture. This is likely to be, for example, in semi-arid areas. Social stresses can be expected to increase and consequent anthropogenic damage to forests may occur.

Policymakers Summary of the Potential Impacts of Climate Change, IPCC Working Group II Report, June 1990

temperature charts for the past decade, a close correlation was found between years of high water temperatures and coral bleaching. If global temperatures continue to rise and the oceans heat up, the worldwide economic consequences of the death of coral reefs would be huge—for example, they currently provide 12% of the global fisheries catch per year.

Wildlife would also be affected by global warming. During a period of gradual climate change, as has happened in the past, herds of grazing animals could move along with the shifting vegetation and were accompanied by the carnivores which preyed upon them. But with rapid climate change, there is no hope of such gradual adjustment. Moreover, today's major wildlife habitats are increasingly hemmed in by farming areas. For example, if the climate starts to change in the Serengeti plain, one of the main sources of tourist income for Kenya, there will be nowhere for the wildlife to go. There will simply be increased competition between them and humans for the resources of the surrounding areas.

A new ecological equilibrium could eventually be restored once the process of rapid climate change ended. But this would take centuries. In the interim, the losses would almost certainly far outweigh the gains, and much of the world's biological inheritance would be gone forever.

Social and political impacts

Although the details of the likely impacts of global warming are still sketchy, they suggest a world system subject to change and increased climatic stress. Moreover, many of the areas which are facing the prospects of an increasingly hostile climate are already among the most vulnerable.

Adapting to climatic change is possible but it will consume considerable amounts of resources which might otherwise be used to meet other pressing needs. Populations which are already restless and disappointed about the progress of national development will become even more dissatisfied if government expenditure has to be diverted to building coastal defences or other such activities. As the prospects of real improvements in living standards recede, political violence and instability may increase.

Extreme climatic events can also have immense social costs. In 1982 Hurricane Isaac destroyed over half the housing and over half the agricultural production in Tonga; and in 1988 Hurricane Gilbert caused damage estimated at over US$870 million in Jamaica.

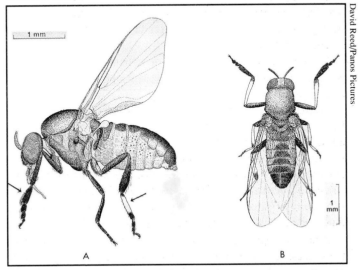

David Reed/Panos Pictures

The health implications of global warming could be deadly. Hotter, wetter conditions could mean greater multiplication of certain disease-carrying insects, such as the "Oncho" fly which spreads river blindness.

Throughout the world, the most vulnerable populations are farmers engaged in subsistence agriculture, residents of coastal lowlands and islands, populations in semi-arid grasslands and the urban poor in slums, in shanty towns, especially in megacities.

A greater number of heatwaves could increase the risk of excess mortality. Increased heat stress in summer is likely to increase heat-related deaths and illnesses. Generally, the increase in heat-related deaths would be likely to exceed the number of deaths avoided by reduced severe cold in winter. Global warming and stratospheric ozone depletion appear likely to worsen air pollution conditions, especially in many heavily populated and polluted urban areas.

Policymakers Summary of the Potential Impacts of Climate Change, IPCC Working Group II Report, June 1990

In those areas which developed a warmer and wetter climate, faster reproduction and greater survival of disease-causing viruses, bacteria and parasites would be encouraged. The incidence of hookworm, schistosomiasis, polio, hepatitis-B and other afflictions would tend to increase. They would also tend to spread into areas where they are now rare or unknown.

The problem of "environmental refugees" will become greater if climatic changes begin to tip fragile productive systems into collapse. Where people find their farms can no longer keep them alive, they have the stark choice of moving or dying. Political frontiers are no barrier against desperate people.

Nitin Rai/Panos Pictures

Increased drought is one of the most serious potential results of global warming. It could create millions of environmental refugees.

Reduction in rainfall in the catchment areas of large rivers would reduce river flows, increasing tension between nations in certain areas. The Blue Nile, for example, rises in Ethiopia, flows through Sudan and then into Egypt. All three countries depend heavily upon the water of the Nile; any major reduction in its flow would make it difficult to agree on equitable arrangements for sharing the shrinking resource.

The range of uncertainty

The potential impacts of global warming are huge and of enormous consequence for all human society. But the range of uncertainty in the scientific analysis to date means that it is still impossible to say exactly what the impact, even at a global level, is likely to be, or when it will be felt.

Taking just the IPCC analysis, and leaving out the more gloomy or optimistic scenarios suggested by others, policy makers are still faced with an extremely wide range of choice. This is illustrated by the two scenarios below.

The IPCC "optimistic" scenario

At the lower, or optimistic, end of the IPCC range of predictions, the sensitivity of the global climate to increased greenhouse gases is low. Under these conditions, and a business-as-usual continuation of trends, the global temperature increase in 2030 is 0.5°C and the sea level rise is 5 cm. The committed temperature rise by the turn of the twenty-first century would be 1.5°C and the sea level rise would be 45 cm. If this prediction is correct, there will be little or no detectable climate change over the next 30 or 40 years. The effects will only begin to become clearly apparent in the second half of the next century.

The IPCC "pessimistic" scenario

At the upper, or pessimistic, end of the IPCC range of predictions, the climate sensitivity to increased greenhouse gas concentrations is high. Under these conditions, and a business-as-usual scenario, the global temperature increase in 2030 is 1.5°C and the sea level rise is 45 cm. The committed temperature rise by the turn of the century is 4.5°C and the sea level rise is one metre.

In this case, the predicted temperature rise up to the year 2030 is roughly three times as high as that over the past century, about 0.4°C per decade. The sea level rise is over 10 cm per decade, 10 times more rapid than that during the last hundred years.

If this is the way the world is heading, the effects will become clearly evident within the next 10-15 years. Major changes in weather patterns, and other implications of global warming, as sketched in this chapter, will be well under way in the 2020s, when today's small children will only be in their thirties.

CHAPTER FOUR

Reducing the Risks

The final details of the debate on global warming are not going to be settled for at least 10 years—possibly much longer. But the human race does not have to be a passive spectator at the drama which has such profound implications for its future. There is an alternative. Greenhouse gas emissions need not follow the business-as-usual trends of the past. A variety of actions can be taken to reduce the risks of global warming. This chapter examines the range of responses available.

Stabilising the greenhouse gas concentration

To eliminate the threat of global warming completely, the concentration of greenhouse gases should be reduced to pre-industrial levels, a goal which is now impossible to achieve. The IPCC calculated the cuts which would be required to hold the concentration at its present levels. These are given in Table 4.1 and

Table 4.1 Cuts in emissions required to stabilise greenhouse gas concentrations at present levels

Greenhouse gas	Required % cut in emissions
Carbon dioxide	60
Methane 1	5-20
Nitrous oxide	70-80
CFC-11	70-75

Source: *Policymakers Summary of the Scientific Assessment of Climate Change,* IPCC Working Group I Report, June 1990

show how drastic the reductions would need to be. Carbon dioxide emissions, for example, would have to drop by 60%, implying a virtual halving of fossil fuel use in transport, industry and electricity generation at a global level.

A scenario, based on research for Greenpeace International, proposed by Dr Mick Kelly, one of the scientists in the Climatic Research Unit at the University of East Anglia in the UK, is designed to bring about a stabilisation of the greenhouse gas concentration at just above today's level by the year 2030. Again, it requires radical change. Among its key features are the following:

• elimination of CFC production by 1995 and avoidance of substitutes which have a greenhouse effect;

• halting deforestation by the year 2000, followed by extensive reforestation;

• reduction in carbon dioxide emissions from fossil fuels to 30% of the present level by 2020;

• a reduction in the annual rise in methane and nitrous oxide concentrations to 25% of the present value.

Even all these changes would not completely eliminate the global warming threat. Under this scenario, the predicted global temperature rise in 2050 would be in the range 0.4-0.9°C above today's temperature, with a final total rise in the range 0.6-1.4°C. But given the relatively slow pace of change, the impact of even the upper figures in these ranges would probably be within generally acceptable limits.

Available measures to reduce greenhouse gas emissions

There is no problem in identifying actions which could be taken to reduce the emissions of greenhouse gases. In most cases, these would be worth taking, ecologically and economically, even if there were no question of global warming. The fact that there is, makes them doubly attractive and means that taking such actions can be seen as what has been neatly defined as a "no regrets" strategy. The situation was summed up in October 1990 by Australia's Minister for Science, Barry Jones: "If we take action and disaster is averted, there will be massive avoidance of human suffering. If we take action and there is no problem, little is lost and we benefit from a cleaner environment. If we take no action and disaster occurs, there will be global tragedy. If we take no action and there is no disaster, the outcome will be due to luck alone [1]."

Energy conservation

Many people fear that energy conservation means a reduction in living standards. This is a complete fallacy. The main point about a high proportion of immediately available energy conservation measures is that they do not call for any sacrifices of living standards. Indeed the reverse is the case. Conservation brings lower running costs for households, businesses and industries; lower energy imports and lower investment in energy supply systems for countries; and lower pollution and greenhouse gas emissions for humanity. It is one of the most economically efficient areas of investment for both the industrial and developing nations.

The present efficiency of energy use throughout the world is low in comparison with what it could easily be. Virtually all the experts agree that there is huge potential for energy saving, which is far more economical than finding new energy sources. Improved efficiency, or conservation, is often described as the cheapest source of energy now available. Most of the potential for saving is in the industrial world, where most of the world's energy is consumed. There are also substantial opportunitities in the industrial, commercial and upper-income household sectors in developing countries, but as the survey on electricity conservation in Brazil (on p 88) shows, it is not just a question of replicating policies and technologies used in the industrial world, since a variety of different factors may be inhibiting conservation in the Third World.

Large amounts of fuel can be saved in buildings. Glass-walled office blocks in tropical cities act as gigantic greenhouses, costing their owners and tenants a fortune in cooling and air conditioning. These mistakes need not be repeated. New buildings can easily be designed to reduce their heat gains and hence the need for cooling; Third World architects and designers can look to traditional methods of designing with nature and use the natural cooling systems of their own vernacular architecture. Careful control of heating and cooling systems and measures to "retrofit" buildings with energy-saving devices can greatly reduce consumption in existing buildings. Greater energy efficiency in buildings pays for itself dozens of times over.

Electricity consumption for lighting can be cut drastically by the use of improved light bulbs. An 18 watt compact fluorescent bulb which can be used in an ordinary lamp socket has a light output equivalent to that of a conventional 75 watt bulb. During its lifetime of about 10,000 hours it saves the emission of more than half a tonne of carbon dioxide.

Transport accounts for about a third of the world's total petroleum fuel consumption. There are now about 500 million vehicles on the world's roads, of which about 400 million are cars. The whole transport sector is crying out for improvements in efficiency. Large US "gas-guzzlers" have fuel consumption figures two or three times higher than compact Japanese or European cars. Taxes and import regulations to keep out the more wasteful models can help reduce carbon dioxide emissions while helping Third World countries to lessen the burden of petroleum fuel imports.

Major savings are also possible in industry. Simple "good housekeeping" measures such as insulating pipes, plugging steam leaks and avoiding waste can often produce dramatic savings for small investments of time or equipment. Significant savings in domestic energy use can also be made by insisting that appliances

An Alternative Approach

One example from a developing country of an alternative approach to planning the electricity sector is the report *Environmentally Sound Energy Development: A case study of electricity for Karnataka State,* produced by the Indian Institute of Science, Bangalore. The official plan for Karnataka calls for annual expenditure for the electricity sector of US$3.3 billion per year, with a particular focus on coal-fired electricity generation. The alternative scenario proposed by the Indian Institute of Science included a range of efficiency-improvement measures, the introduction of solar water heating and LPG stoves, cogeneration in sugar factories and small-scale hydro systems, and decentralised rural power generation based initially on biomass and later on solar power. This strategy will cost only US$0.6 billion per year and produce a negligible increase—400 tonnes—of carbon emissions compared to the 830,000 tonnes per year in the official plan [2].

Bernard Régent/Hutchison Library

In the wealthy OECD countries, 54% of energy-related nitrous oxide emissions come from transport.

meet minimum energy efficiency standards. Electricity tariffs which charge a punitive rate for consumption above a certain level ~uld cut down on wasteful use of air-conditioning.

~~re of the potential scope for energy conservation, at least
~rovided by the study *Energy for a Sustainable*
~~d out by four well-known energy analysts,
Amulya Reddy of India, Thomas
~bert Williams of the United States.

nd-use approach" in which, rather than
.on trends into the future, the focus was
.d the best means of satisfying them. In its
~vas "especially interested in understanding

how patterns of energy use might be shaped so as to promote the achievement of certain basic societal goals—equity, economic efficiency, environmental soundness, long-term viability, self-reliance and peace."

The results of the analysis were startling. They showed that it appeared to be technically and economically feasible to provide a much higher standard of living for a doubled world population in the year 2020 with only a 10% increase in energy consumption over the 1980 level. The scenario envisages an increase in Third World energy consumption sufficient to provide a level of energy services equivalent to that in Western Europe. In the industrial world, energy consumption falls by 50%, though there is still substantial economic growth.

Demonstrating the technical and economic feasibility of such a low energy future does not, of course, mean that the policy choices necessary for its implementation will necessarily be made. But it clearly shows what is possible. As the study says: "Contrary to widely held beliefs, the future for energy is very much more a matter of choice than of destiny."

There are also some hopeful signs that the potential of energy conservation is being realised by some of the industrial countries. Between 1973 and 1985, Japan's gross domestic product (GDP) grew by 46% while energy consumption scarcely changed. This represented an improvement in the energy efficiency of economic

---

The multilateral development banks' overwhelming influence on energy and economic development in developing countries should be used to promote a shift to [a] more efficient and sustainable energy path. The first steps down that path can be taken nearly immediately, with technical assistance for institution-building in developing countries, demonstrations of energy conservation projects, and improvements in the ability of the banks' lending staff to design and administer loans for energy-efficiency....Without this redirection of the banks' efforts, many developing countries will never escape the capital shortage and environmental damage that constrain their progress. There is no longer an alternative.

The Least-Cost Energy Path for Developing Countries: Energy-efficient inve *for the multilateral development banks, 1991, from an executive summary b* Philips, International Institute for Energy Conservation, United States.

output by 34%. The story is similar in Denmark where the present energy demand per unit of GDP is less than 70% of its 1972 level.

Elimination of CFCs

Under the revised Montreal Protocol, the manufacture and use of CFCs is due to be banned completely by the year 2000. There are, however, no major technical reasons why they could not be replaced immediately in most cases. Products which can replace these CFCs and do not damage the ozone layer are already available. Prolonging the use of CFCs for almost another decade means a needless addition of these potent and damaging gases to the atmosphere.

The threat posed by the large quantities of CFCs already incorporated in refrigerators and air-conditioning units must also be kept in mind. These need to be recycled or captured for destruction when the equipment is being repaired or scrapped. Unless this is done, there will be a continuing release of these stocks of CFCs over the next few decades.

Elimination of CFCs is also necessary for the preservation of the ozone layer. The fact that these gases, on a business-as-usual projection, would contribute 20% to the total greenhouse effect by 2030, is just another compelling reason for taking immediate action to eliminate their use completely.

Even if an immediate ban on the use of CFCs was successfully imposed, fridge dumps like this will continue to release CFCs into the atmosphere for several decades.

Switching fuels

The greenhouse gas emissions from use of the different fossil fuels vary considerably. For the same amount of heat produced or electricity generated, natural gas produces about 40% less carbon dioxide than coal and about 25% less than oil. Switching between fuel sources can therefore bring substantial savings in carbon dioxide emissions.

Fuel switching, however, should not be considered in isolation, but as part of an overall energy strategy for the minimisation of carbon dioxide emissions. Any decision to use natural gas for electricity generation, for example, should always be balanced against the much lower efficiency of this compared with using the gas directly for cooking or industrial heating. If natural gas is used in a power station to produce electricity which is then used for cooking, the overall efficiency of energy use is likely to be less than 25%. The efficiency of using the gas directly in a boiler or gas cooker may be up to 75%. Using the natural gas for power generation, in that case, will produce carbon dioxide emissions which are three times higher than using it directly.

The difficulties must also be recognised. In many developing countries there are real problems about exploiting sources of gas, transporting it to potential markets and supplying those markets, whether industrial or domestic, with the appropriate appliances. To encourage domestic consumers, for example, millions of gas cookers would be needed.

Where coal is the only practical choice of fuel for electricity generation, power stations should be as efficient as possible. Efforts should also be made to use the waste heat, thus cutting down on fuel needs elsewhere.

Nuclear power stations do not emit carbon dioxide when they are generating electricity. At present, nuclear power saves the emission of about 450 million tonnes of carbon per year. Some people feel that it can play an increasingly important role in combating the global warming threat. However, there are other major problems associated with nuclear power: radioactive pollution, high cost and low public acceptability.

Moreover, even if all its problems could be overcome, nuclear power is not "the answer" to the problem of carbon dioxide emissions. Its only practical use is in electricity generation which accounts for about 35% of total fossil fuel consumption. So even a massive expansion of nuclear power would leave the bulk of carbon dioxide emissions untouched.

Reducing methane and nitrous oxide emissions

The methane produced from rice paddies is a result of the breakdown of organic material in the soil when it is under water. Changes in irrigation practices, with longer dry periods, varieties of rice producing less residue and shorter growing periods could all reduce emissions. But considerably more research is needed before a workable strategy can be devised.

Part of this effort is already under way. The International Rice Research Institute (IRRI) in the Philippines announced the launch of a major research programme at the end of 1990. This will examine the contribution of rice-growing to global warming, methods of reducing this impact, and possible ways of adapting rice production to climate change.

Neil Cooper/Panos Pictures

Doubts remain over the extent to which rice paddies are responsible for the production of greenhouse gases. More research needs to be done, not just in the North, but in those developing countries where rice is heavily cultivated.

Even less is known about the overall quantities and the distribution of methane emitted by livestock. All that is certain is that the emissions represent food energy which has not been turned into milk or meat. Improved feeding practices which would reduce methane emissions could almost certainly be devised; it might also be possible to breed stock selectively for lower methane production. The possibilities of introducing such measures among nomadic herders or poor farmers are limited. But where stock are stall-fed, the range of options is wider.

So little is known about the origins of the nitrous oxide building up in the atmosphere that effective strategies for its limitation cannot yet be formulated. The excessive use of nitrogenous fertilisers is one major cause of emissions which should be restricted for a variety of other good reasons. Another significant source appears to be forest burning, which it is also desirable to limit on other grounds.

Biomass fuels and cooking stoves

Biomass fuels are those such as wood or crop residues, which are grown. They can be used on a sustainable basis, with an equivalent amount to that used being replanted. When this is done, there are no net carbon dioxide emissions because the growing crops use as much carbon dioxide as is released when the fuel is burnt. If the energy being produced is substituting for fossil fuels, there is a corresponding saving in carbon dioxide.

Biomass fuels are already used in a sustainable way in a variety of rural industries in the developing world. Sugar and rice mills, oil palm factories, and other agro-industries regularly rely on their own waste products to provide the energy they need. Sawmills often use their own offcuts and waste for power generation and to produce heat for timber drying. Enterprises of this type should be encouraged to substitute biomass fuels for fossil fuels to the maximum extent economically and technically feasible. In some cases, factories may be able to provide electricity to local communities or feed surplus supplies into the grid.

But the use of biomass fuels is not an easy or universally applicable option. Care needs to taken to ensure that biomass fuel projects are technically and economically sound if they are not to add to the problems of the developing world. Too often, biomass energy projects sponsored by external donor agencies have turned out to be expensive failures consuming technical and financial resources the recipient countries could ill afford. For example, the attempt by the Philippines to develop a dendrothermal programme in which electricity was to be produced in wood-fired power stations fuelled from special plantations of fast-growing trees, has turned out to be a major disappointment. The programme foundered as a result of problems with the plant and the failure of a high proportion of the tree plantations.

Also, when biomass fuels are not used on a sustainable basis, there is no saving in carbon dioxide emissions. Wood emits about 35% more carbon dioxide than coal and, if it is not burnt efficiently, can also emit a certain amount of methane. Charcoal-making is particularly damaging since it produces a substantial amount of methane. It is also notoriously destructive of forest resources, since charcoal-makers cut down whole trees.

Improved cooking stove programmes, especially those focused upon the dissemination of improved charcoal stoves in urban areas, can therefore play a useful role in helping combat global warming. By lowering charcoal use, they reduce both carbon dioxide and methane emissions.

The question of whether the gradual switch away from traditional wood or charcoal stoves to kerosene or liquid petroleum gas (LPG) stoves increases or reduces greenhouse gases is relevant in this context. The answer is that if the woodfuels are being supplied on a sustainable basis, which is rarely the case, the switch to fossil fuels will increase carbon dioxide emissions. Otherwise, the shift to fossil fuels will bring about a significant saving in carbon dioxide emissions. This is because of the considerably higher efficiency of the fossil fuel stoves. A well designed wood or charcoal stove may have an efficiency of 15-20%, whereas a kerosene or LPG stove is likely to be two or three times more efficient. For the same cooking task, the woodstove will therefore use two to three times more energy than the fossil fuel stoves, even before the higher emissions per unit energy from wood and the methane emissions from charcoal kilns are considered.

Renewable energy technologies

The use of most renewable energy sources does not produce any carbon dioxide emissions. Increasing the energy supplied from renewable sources is therefore seen by many as a major element in any strategy to reduce carbon dioxide emissions. So far, however, the contribution of renewable energy sources, apart from hydro power, to world energy supplies is extremely small.

The saving in carbon dioxide emissions as a result of the present generation of electricity from hydro power is about 530 million tonnes of carbon, about 20% more than nuclear power. Large-scale hydro plants are, however, now largely out of favour. The financial costs are high and, in the developing world, the economic returns have generally been slow in materialising. The environmental costs are also heavy. The creation of storage lakes behind dams, in addition to flooding large areas of land and displacing farming communities, can lead to a range of social and ecological problems.

Small-scale hydro schemes avoid these problems since they do not impound such large volumes of water. Their output, however, tends to be small, matched to the consumption levels of villages rather than the urban and industrial areas where the vast bulk of electricity is consumed. The number of suitable sites is also limited. Except in a few favoured countries such as Nepal with its many waterfalls, the potential contribution of small-scale hydro to electricity supplies at a national level is generally low.

Solar energy can also be used to reduce the consumption of fossil fuels. Solar water heaters are already widely commercialised and are in common use in countries like Japan, Australia and Israel. They could be more widely used, especially if governments were to provide tax allowances or other incentives for home-owners and businesses to install them.

Solar photovoltaic systems are used to produce small quantities of electricity for areas where there is no electricity grid. Their use can certainly be extended. But they are very costly and their impact on overall energy consumption is likely to be small. At present, a photovoltaic system producing enough electricity for a 1 kilowatt hot plate would cost about US$20,000. Costs are falling, but the contribution of photovoltaic systems to national energy supply systems is unlikely to be significant over the next few decades.

Wind power has made considerable progress in the past two decades. In California, there is now a substantial wind power industry with about 15,000 machines, grouped in "wind farms" supplying power to the electricity grid. The total output of all the machines is about 1,350 megawatts (MW). Denmark, the UK, the Netherlands and other countries are also encouraging the use of wind power. India has established a prototype wind farm in collaboration with Denmark and is planning to begin the manufacture of wind turbines.

It is estimated that the total installed wind power capacity in the world by the year 2000 will be about 9,000 MW. While this is impressive, and augurs well for the longer term, it also needs to be seen in proportion; a single large coal-fired power station has a capacity of around 2,000 MW. However, some energy experts predict a rapidly accelerating expansion of wind power capacity from the first decade of the twenty-first century.

Another highly beneficial technology which deserves mention among the renewables is the capture of methane from large urban waste dumps. This has proved to be an economic activity in some areas and is highly beneficial from the greenhouse point of view. Instead of being released to the atmosphere, the methane is used to provide energy for electricity generation or heating boilers, thus replacing an equivalent amount of fossil fuel.

In summary, renewable energy sources help to reduce the level of greenhouse gas emissions and deserve to be encouraged whenever possible. But it must also be recognised that they provide no instant or easy solution.

Reforestation

The amount of carbon absorbed by a growing tree depends upon the species, the climate and the soil; it also varies with the age of the tree. The rate of growth, and hence the amount of carbon fixed, or sequestered, is greatest in the early years and tapers off as the tree reaches maturity. As a very rough guide, it can be said that a growing forest fixes about 10 tonnes of carbon per hectare per year.

This means that to absorb 10% of the present carbon dioxide emissions to the atmosphere, it would be necessary to plant an area of about 700,000 sq km, roughly the area of Zambia or Turkey. To absorb all the annual emissions would require planting an area the size of Australia. Such a plantation would act as a carbon absorber for perhaps 30-40 years after which it would be a carbon store for as long as it remained standing.

Tree growing, even on the most massive scale likely to be feasible, cannot therefore fully compensate for the rate at which greenhouse gases are being added to the atmosphere. Nor does it provide a permanent solution, as the carbon fixed in the trees will be released again as they decay or are burned. Nevertheless, a commitment to increased tree growing by every country would slow the build-up of greenhouse gases, and if the wood is subsequently used on a sustainable basis (with replanting equalling use) there will be a decrease in fossil fuel consumption and so in emissions.

Exotic possibilities

A number of exotic possibilities have been suggested for reducing the threat of global warming. None is likely to see use within the next few decades.

The fact that volcanic eruptions, which throw large amounts of dust in the air, have been known to cause a marked global cooling has led some scientists to wonder whether the same effect could be achieved deliberately. The suggestion is that it might be done by large fleets of aeroplanes spraying sulphur dioxide into the atmosphere. The costs, and potential side-effects, of such an exercise make it a highly improbable option.

It has also been suggested that carbon dioxide might be filtered out of power station exhaust gases before they are released into the atmosphere. This is technically feasible and is done on a small scale as a means of obtaining carbon dioxide for commercial uses. The problem with carrying it out on a large scale, apart from the cost, is the disposal of the carbon dioxide, since the quantities are far in excess of any potential uses. It would be possible, in principle, to liquefy it and pump it down into the deep ocean. But this is unlikely to be feasible economically or practically. Indeed, the whole process could consume most of the energy produced by the power stations.

Another suggestion is to seed certain parts of the oceans with iron dust or filings. The availability of iron is the limiting factor governing the number of algae living in certain ocean areas. By making up the iron deficiency, greater numbers of algae would grow, absorbing carbon dioxide as they did so and carrying it to the bottom with them when they died. The sheer scale of the effort required to have a significant impact, plus the highly unpredictable environmental consequences, make this also an unlikely option.

Progress to date

Discussion on ways to restrict the growth of greenhouse gas concentrations has been under way at an international level since the 1988 Toronto Conference on The Changing Atmosphere. This called for a 20% reduction in the 1988 level of carbon dioxide emissions by the year 2005.

Since then, the question of targets has generated a considerable degree of international acrimony. The United States, which accounts for a quarter of the world's emissions, is adamantly opposed to them; so also are Saudi Arabia and the USSR. Japan has also objected on the grounds that it has already achieved a high degree of energy efficiency and that it would be unfair to expect it to match the cuts in emissions which could so easily be made by more wasteful nations.

However, a number of countries have committed themselves. The German cabinet has, for example, voted for a 25% reduction in carbon dioxide emissions below 1987 levels by the year 2005. The Netherlands is aiming to stabilise emissions by 1995 and to reduce them by 8% by the year 2000. Norway has pledged to stabilise consumption at present levels. The UK has stated its intention to stabilise emissions at present levels by 2005. Overall, however, European Community (EC) member states have not been able to agree on targets and in September 1991 the EC proposed a new plan involving taxes on fuels which reflect the carbon they emit when burnt and the energy they produce.

Carbon taxes and tradeable permits

Price is one of the most important factors determining which fuel people choose and how much they consume. Economists have suggested that fuel prices could be adjusted by means of "carbon taxes", as a way of combating global warming. The carbon tax would be levied on the fuel in proportion to the amount of carbon dioxide it emitted. Under this scheme, coal would bear a higher tax than petroleum fuels, with natural gas bearing the lowest burden.

While energy pricing is an important factor in determining its efficient use, taxes of this type have practical limitations, particularly in developing countries. Energy is not a particularly free market, and most consumers do not have much choice about which fuel they choose; in the case of electricity they have no say over what fuel is used to generate it. Carbon taxes could especially affect the poor, who tend to have least flexibility in choice of fuel.

Another idea being promoted by economists is the use of "tradeable permits" in carbon dioxide emissions. These permits would allow a country or an organisation to emit a certain amount of carbon dioxide. The total global level of emissions would be decided by an international body such as the UN, which would allocate a permit to emit a certain amount to each country. Within countries, the permits would be shared between fuel users.

Any country which wanted to emit more than its allocated share of carbon dioxide would have to purchase a corresponding allocation from a country which was not using its share to the full. Organisations such as power utilities would also be able to sell and purchase permits between themselves. The result, in theory, would be to minimise the overall economic cost of achieving a given level of carbon dioxide emissions. Any country which did not conform would be subject to sanctions of some kind, such as a ban on exports.

The difficulties facing any such programme are obvious and probably insurmountable in the immediate future. It would be extremely difficult to monitor carbon dioxide emissions accurately at a national level, or to devise a workable system to avoid cheating. Neither is there any immediate prospect of the UN or any other international organisation being trusted to allocate permits fairly—between Iraq, Saudi Arabia, the United States, and India for example—or to identify and deal with violations.

Nevertheless, if global warming turns out to be as severe as some fear, so that the global community is forced into collaboration, some system of setting limits to emissions and sharing the burden of staying within them will undoubtedly be required. Tradeable permits are one possible approach.

Meeting the costs

Most of the immediate steps to mitigate the threat of global warming bring net benefits. Simple and effective energy conservation measures in industries, businesses and households typically have payback periods of two or three years, sometimes much less. Initial costs, however, often have to be incurred before savings can be realised. This can be a major deterrent. Government intervention in the form of tax incentives or grants may therefore be required to get effective and large energy conservation programmes under way. Given the many calls on their resources and the understandable reluctance of people to pay higher taxes, many governments feel they cannot afford such expenditures and resist becoming involved in energy conservation. However, it is instructive to look at what governments are prepared to spend to lessen the threat which they feel other nations pose to their national and political security.

Spending on armaments

The Swedish International Peace Research Institute (SIPRI) has monitored expenditure on armaments for the past few decades. Governments are naturally shy of disclosing how much they spend on weapons. Nevertheless, the SIPRI researchers are adept at

Table 4.2 Estimated total military expenditure by selected countries in 1989 (US$ million at 1988 prices)

Country	Expenditure	Country	Expenditure
United States	289,000	Thailand	2,100
(EC total)	(152,000)	Ethiopia	780
USSR	150,000*	Philippines	700
France	36,000	Peru	620
FR Germany	35,000	Zimbabwe	390
UK	34,000	Nicaragua	348
Japan	29,000	Bangladesh	311
Italy	20,000	Sudan	270
China	11,000*	Kenya	220
India	9,000	Sri Lanka	205
Brazil	3,700	Tanzania	170
Pakistan	2,800	Costa Rica	19
* Very rough estimate			

Source: Based on data in *SIPRI Yearbook 1990: World Armaments and Disarmaments*, OUP, Oxford, 1990

winkling out the facts. Table 4.2 shows estimates for expenditures by a selection of industrial and developing countries for 1989.

The amounts are huge and show clearly that all the major industrial countries are perfectly capable of diverting substantial funds from military budgets to the more positive and urgent task of preserving the equilibrium of the global atmosphere. What are perhaps more surprising are the sums of money spent by a significant number of Third World countries. In some cases, military spending is greater than the combined total spent on health and education. It is also estimated that up to one-third of all developing country debt is a result of arms imports.

No regrets

Paying an insurance premium is a widely accepted way of dealing with risk. Indeed, military expenditure is normally justified on the grounds that it is a type of insurance designed to protect a country and its people against threats to their security and well-being.

Global warming is certainly a threat to the security and well-being of the world. Its costs and disruptive effects, especially if the more pessimistic predictions are realised, are comparable to those of a major international war. It is therefore logical to pay an insurance premium to prevent it happening.

The point about insurance is that it is paid under conditions of uncertainty. Once the factory is on fire or the ship has sunk, it is too late to get insurance; at that stage, all that can be done is to face the consequences of the catastrophe. Similarly with global warming; once it is clear that it is happening, it is too late to prevent it.

As this chapter shows, a great deal can be done to reduce the threat of global warming. Taking these measures is a prudent form of insurance. The fact that many are economically and environmentally justified in their own right means that implementing them can be part of a "no regrets" policy. Even if the threat of global warming turns out to be much less than feared, it will not harm the world if the ozone layer is protected, energy is used more efficiently, acid rain and smog are reduced, and more trees are grown.

Acting against the global warming threat in this way is a rare instance in which the world can have its cake and eat it. It is an insurance policy from which, whatever happens, the returns are greater than costs.

Jiri Polacek/Panos Pictures

The effect of acid rain in the Czech and Slovak Republic. Up to a million hectares of the country's forests could be dying. Any actions to reduce pollution and improve current patterns of energy use will be of benefit to the environment, whatever the reality of global warming.

Climate change negotiations

A group of 120 countries is attempting to complete negotiations for a framework convention on climate change in time for signing at the "Earth Summit" in Rio de Janeiro, Brazil, in June 1992.

The convention will not succeed in setting binding targets for limiting greenhouse gas emissions, but will probably establish the objective of a "safe level"; will refer in general terms to the need for stabilising emissions in industrialised countries while allowing for expansion of developing country outputs; and will touch on the issues of incentives for energy efficiency and provision of finance for the use of climate-friendly technologies in the Third World. The creation of a new fund for this purpose is under discussion.

The aim is not to solve the problem but to begin a process, which is why it is termed a "framework convention".

Negotiations have proceeded on a North-North axis—with the industrialised countries trying to thrash out a common position on stabilisation of emissions—and on a North-South basis, to work out the terms on which developing countries will contribute to the overall strategy. Brazil, China, India, Malaysia, Mexico and Saudi Arabia have been the most active Southern negotiators.

US unwillingness to set targets has been a major stumbling block, mainly because, as a study published by Britain's Royal Institute of International Affairs commented, "The US economy is as dependent on fossil fuels as a heroin addict is on the needle." Washington has favoured a general request for countries to reduce output of greenhouse gases plus an increase in the size of "sinks", such as forests. Sinks will be dealt with in the convention, but less specifically than emissions.

The fourth negotiating session, in Geneva in December 1991, involving 120 countries, is being followed by tough bargaining in smaller groups—and "high-level political decisions in capitals will undoubtedly be needed to seal the bargain in 1992," according to Jean Ripert, chairman of the Intergovernmental Negotiating Committee.

CHAPTER FIVE

Policy Implications for the Third World

Many people in the Third World legitimately feel that global warming is a problem which has been caused by the industrial countries. They see no reason why the responsibility for resolving it should not rest primarily with those who caused it, and who are still responsible for three-quarters of the greenhouse gases currently being added to the atmosphere.

The world's atmosphere, however, is a single entity. The impact of global warming, if it occurs, will be felt by everyone, whether they caused it or not. Indeed, it may be that the Third World will suffer some of the worst effects.

In pursuing their development strategies, Third World countries therefore have to take a global as well as a national view. They must respond to the urgent developmental needs of their people. They clearly cannot afford to accelerate a process of which they themselves are likely to be the principal victims, yet responding to the possibility of climate change takes resources which may be already insufficient to meet more immediate demands. The threat of global warming thus poses a series of policy questions, not just for Third World governments, but also for development banks and agencies which provide investment capital and technical assistance for development projects with global warming implications.

Allocating responsibilities

There is no doubt that the industrial world has the major responsibility for historic and present carbon dioxide emissions

The industrial world bears the major responsibility for past and present carbon dioxide emissions.

from the use of fossil fuels. But the origin of the remainder of the greenhouse gases and the way in which overall emissions are likely to evolve in the future are more contentious matters.

The argument is important. As the world gradually comes to grips with the need to do something about global warming, logic and fairness demand that those who do the most damage should make the biggest efforts. That is why there is now an acrimonious debate about the present and likely future rankings in greenhouse gas emissions.

Distribution of fossil fuel emissions

The approximate division of 1990 fossil fuel consumption between the main consuming countries and regions is given in Table 5.1. It shows that over a quarter of the consumption is in North America; this is despite the fact that it contains only about 6% of the world's population. The USSR and Eastern Europe come next with about 22%. Western Europe's share is 16.4%. China, with about a fifth of the world's population accounts for 9.2% of fossil fuel consumption, while India's share is just 2.4%.

These proportional shares have been gradually changing and are likely to continue to do so in the future. Energy consumption growth

Table 5.1 Fossil fuel consumption in 1990

Country or region	Fossil fuel consumption (million tonnes of oil equivalent)	Proportion of total (%)
USA	1746	24.8
Canada	159	2.3
Latin America	357	5.1
Western Europe	1155	16.4
USSR	1246	17.7
Eastern Europe	311	4.4
Middle East	239	3.4
Africa	202	2.9
China	646	9.2
Japan	365	5.2
India	170	2.4
Asia*	337	4.8
Australia/New Zealand	98	1.4
World	7031	100.0

* Excluding China, Japan and India

Source: *British Petroleum Statistical Review of World Energy,* 1991

in the industrial world has slowed considerably in the past two decades and was about 20% in total over the 1980s. Growth rates in the Third World during the same period were much higher. In both Africa and Latin America, energy consumption increased by around 40% in the 1980s, and by 60% in Asia excluding Japan.

But these high growth rates can be highly misleading unless put in their proper context. In each case, the increase started from very low initial figures. Despite the growth, the total consumption of fossil fuels in Africa is still less than 3% of the world total; that in Latin America is just 5%; and India and China together account for just 11.6% [1]. Although it is likely that Third World energy consumption will continue its faster growth rate, it will be a long time before it comes close to the level of carbon dioxide emissions of the industrial world, let alone approaches the level of consumption per person.

Another frequently expressed environmental concern is that coal, which tends to be the cheapest form of energy for power generation and other large-scale uses in some developing countries,

The countries of the industrially developed world are the main source of greenhouse gases and therefore bear the main responsibility to the world community for ensuring that measures are implemented to address the issues posed by climate change. At the same time they must see that the developing nations of the world, whose problems are greatly aggravated by population growth, are assisted and not inhibited in improving their economies and the living conditions of their citizens. This will necessitate a wide range of measures, including a significant increase in energy use in those countries, and compensating reductions in industrialised countries. This transition to a sustainable future will require investments in energy efficiency and non-fossil fuel sources. In order to ensure that these investments occur, the global community must not only halt the current net transfer of resources from developing countries, but actually reverse it.

Declaration of 1988 Toronto Conference on the Changing Atmosphere

is likely to increase its share in Third World consumption. This is true, but again the issue needs to be kept in perspective. The OECD nations and the countries of the former Soviet bloc presently account for 63% of world coal consumption. China's consumption is 24% and India, the next largest consumer in the developing world, uses a mere 4%. Carbon dioxide emissions from coal are primarily a problem of the industrial world, and are likely to remain so for a considerable time.

Distribution of other emissions

Apart from the CFCs which are primarily emitted by the industrial world, great uncertainty surrounds the sources of other greenhouse gas emissions. Any allocation of responsibility on a national basis can therefore be made only on an extremely tentative basis.

Estimates of emissions from deforestation and the extension of farming are in the range 1-2 billion tonnes, but no one can be certain. Many of the estimates for Third World emissions appear to be based on figures for rates of deforestation which are now acknowledged to be too high and are being revised downwards.

It is also highly probable that the rates of carbon dioxide emission assumed to result from deforestation exaggerate the true effect. Much of what is defined as deforestation by foresters does not involve the complete destruction and combustion of all the existing vegetation. When traditional cultivators move into forested areas, the forest structure is undoubtedly destroyed, but a considerable proportion of the vegetation remains or quickly regrows.

There are similar major question marks over the origins of methane emissions. Estimates of emissions from livestock and rice growing are highly speculative. Little is known in detail about the undoubtedly large amounts released from coal mining as well as from petroleum exploration and production. In countries with natural gas networks there are also emissions from pipelines and appliances, but again the statistics are unavailable or not reliable.

Least certain of all is the position over nitrous oxide emissions. That they come from the soil and from fossil fuel and wood burning is accepted; but how the total amounts emitted are allocated between these sources and different regions is still largely unknown.

Blaming others

The World Resources Institute (WRI), in Washington DC (US), sparked off a major controversy in the middle of 1990 when it published a world league table of responsibility for greenhouse gas emissions. This clearly ranked the United States in first position followed by the USSR, but controversially allocated the next three places to Brazil, China and India. Commenting on its presentation, the WRI stated:

A major finding of *World Resources 1990-91* is that Brazil is a larger source of carbon dioxide emissions than the United States in 1987, primarily because of massive deforestation. China's greenhouse gases, 6.6% of the world's total, stemmed from fossil fuel use and rice growing. India's extraordinarily high methane emissions—98,000 metric tonnes, second only to the United States—are traceable to rice growing and livestock.

This was immediately seen by Third World groups as a means of shifting the focus away from the massive releases of carbon dioxide and CFCs by the United States and other industrial countries. The Centre for Science and Environment (CSE) in New Delhi (India) called the report an "excellent example of environmental colonialism", adding:

The report of the World Resources Institute (WRI), a Washington-based private research group, is based less on science and more on politically motivated and mathematical

juggery. Its main intention seems to be to blame developing
countries for global warming and perpetuate the current global
inequality in the use of the earth's environment and its
resources....The WRI's report is already being quoted widely and
its figures will definitely be used to influence the deliberations
on the proposed, legally binding, global climate convention [2].
The CSE heavily questions the basis of the WRI figures for
deforestation rates, pointing out, for example, that the latest figure
for India is 47,500 hectares per year rather than the 1.5 million
assumed in the WRI calculations. It is also sceptical of the basis for

**Table 5.2 Different approaches to ranking the top 15 greenhouse
gas-emitting countries**

		WRI calculations*		CSE calculations**	
Ranking	Country	Net greenhouse gas emissions[†]	Country	Net greenhouse gas emissions
1	United States	1000	United States	1532
2	USSR	690	Brazil[††]	1017
3	Brazil	610	USSR	730
4	China	380	Canada	252
5	India	230	FR Germany	155
6	Japan	220	Japan	140
7	FR Germany	160	United Kingdom	132
8	United Kingdom	150	Australia	112
9	Indonesia	140	Saudi Arabia	97
10	France	120	Colombia	86
11	Italy	120	Côte d'Ivoire	82
12	Canada	120	German DR	82
13	Mexico	78	Myanmar	81
14	Myanmar	77	Laos People's Rep.	78
15	Poland	76	Poland	77

* World Resources Institute, Washington DC
**Centre for Science and Environment, New Delhi
[†] Million tonnes of carbon equivalent
[††]When the 1978-88 average figure for Brazilian deforestation is used the net emissions from Brazil fall to 197 million tonnes

Source: Based on a table in *Global Warming in an Unequal World: A case of environmental colonialism*, Centre for Science and Environment, New Delhi, 1991

the WRI's methane emissions, arguing that the figures used have been based on a limited amount of research in the industrial world, which is of highly doubtful applicability to rice cultivation in developing countries.

The CSE, nonetheless, accepted the WRI figures for the sake of argument but recalculated the league table on a radically different basis, which it considers much fairer. The point made by the CSE is that there are natural sinks or chemical processes which absorb or destroy the greenhouse gases, apart from CFCs. Approximately half the present carbon dioxide emissions, for example, are absorbed by the oceans and land vegetation. It is not the emission of greenhouse gases as such which is the cause of increasing greenhouse gas concentrations in the atmosphere, but the fact that the present level of emissions exceeds the capacity of the available natural sinks.

In the CSE calculations, the natural greenhouse gas sinks are allocated on a per head basis to each country. The net contribution of each country to the global warming threat is then the amount by which its total greenhouse gas emissions exceeds its allocation of sinks. The results of this set of calculations are given in Table 5.2. Commenting on its approach, the CSE remarks:

> The WRI report makes no distinction between those countries which have eaten up their ecological capital by exceeding the world's absorptive capacity and those countries which have emitted gases well within the world's cleansing capacity. India, for instance, has been ranked as the fifth largest contributor of greenhouse gases in the world. But compared to its population —16.2% of the world's in 1990—India's total production of carbon dioxide and methane amount to only 6% and 14.4% respectively of the amount that is absorbed by the earth's ecological systems. How can, therefore, India and other such countries be blamed even for a single kilogramme of the filth that is accumulating in the atmosphere on a global scale and threatening the world's people with a climate cataclysm? In fact, India can double its carbon dioxide emissions without threatening the world's climate [3].

As can be seen in Table 5.2, the United States shows up as an even greater contributor to the greenhouse gas increase, while China,

Timber imported from Malaysia to Japan. An important factor in tropical deforestation is the demand for wood by industrialised countries.

India, Indonesia and Mexico drop out of the top 15 emitters. Brazil, on the other hand, moves up to second place with about two-thirds the emissions of the United States. The CSE points out, however, that when Brazil's deforestation rate is taken as the average for the 1978-88 decade rather than the peak rate for 1987 assumed by the WRI, it drops to fifth place with a contribution of just under 13% of that of the United States.

A global responsibility

It is, indeed, important that the burden of dealing with the global warming threat is shared equitably. It is also clear that the problem is a global one which will have to be solved by concerted global action. This will require a degree of all-round responsibility and, indeed, generosity which has not been achieved so far.

Far too often, many developing countries feel, they have been singled out and asked to conform to far more stringent environmental standards than the industrial world sets for itself, to the extent that they are being denied the development opportunities which the industrial world still takes for granted and reserves for itself. Global environmentalism, its critics argue, is thus a way of perpetuating the existing economic divisions of the world. Anil Agarwal, one of India's most prominent environmentalists and director of the CSE, has said:

> ...a person who is living in India like me may feel very strongly about the fact that if the Americans keep on building more power stations based on coal, this leads to global warming, which leads to a sea level rise, and then drowns Bombay and Bangladesh. Do I have the right to say something about it or not? What are the levers of power that I have? None at all. Which means that the moment I accept "global environmentalism" then I accept —given the realities of the world as it exists today—a second class citizenship in this world. I allow the Americans to go to the World Bank and say there will be a green conditionality on Narmada [dam project] but I have no right, no forum and no power whatsoever to say that there will be environmental discipline on energy consumption patterns in the United States of America. What do I do about that? It has tremendous repercussions for me, extraordinary repercussions for my economy. If global warming takes place and the climate changes, how am I going to adjust to that?...If the demand is for global governance in a sense of fair rights to all of us, then let us talk about a right to a clean environment for every citizen in the world, to which every government will have to submit itself [4].

We are now faced with the fact that the total resource base of mankind is in danger. Its limits are in sight and its quality is deteriorating. It matters little who is in control because ultimately the very existence of all will be at stake when no action is taken to halt the depletion of our common resource base. Other developments in the world have underscored the fact that what happens to people in one part of the world might affect the entire world. The structure of human endeavour today is such that the actions of one nation affects the entire structure. Global interdependence is an emerging fact which is not well appreciated. Nevertheless, there is a growing realisation that only through coordinated, interdependent actions can we safeguard our common well-being. Here, obviously, is the central point and rationale for globally sustainable development strategies.

Emil Salim, Indonesian Environment Minister, at the 1988 Toronto Conference on the Changing Atmosphere

If a genuine global effort to combat global warming is to take place, the depth, strength and justification of such statements will have to be recognised. The developing world will have to be provided with the resources which will enable its governments and people to look beyond the urgent pressures of day-to-day survival to the longer-term protection of the environment on which all ultimately depend.

Increasing Third World participation

Consciousness of the need to bring about increased and more informed Third World participation in the global warming debate is growing. The IPCC set up a Special Committee on the Participation of Developing Countries to study the factors which have inhibited Third World involvement in its deliberations. The committee identified five areas of particular concern:

- insufficient information;
- insufficient communication;
- limited human resources;
- institutional difficulties;
- limited financial resources.

Many developing countries do not have sufficient information on the scientific basis of the global warming issue, and why it is causing such widespread concern. Neither do they have the data for informed discussion of the policy options to restrict global warming which are open to them, or of ways of coping with climate change if it takes place as predicted.

Lack of the necessary qualified personnel, and institutional weakness were also found to be major factors inhibiting Third World discussions on global warming and the development of policy responses. The fact that Third World countries are short of the financial and technical resources required for meeting immediate day-to-day budgetary and developmental needs is also a major obstacle to involvement in the global warming issue while it remains at its present level of imprecision.

The committee made a number of recommendations with the specific objective of increasing the involvement of Third World participants in the work of the IPCC. If they are implemented, they will help at the scientific and senior government official level. But it still leaves journalists, opinion formers, and middle-ranking decisionmakers largely outside the mainstream of the debate.

Support for research

Research and technical capabilities are low in many Third World countries. This means it is impossible for them to monitor potential global warming impacts or to study ways of coping with them if they occur. A report by a Commonwealth Group of Experts stated:

> We have argued that there are good grounds for accepting the scientific consensus that rises in global mean temperatures and sea level are likely. Beyond this, the speculative nature of the regional details of climate change is unsatisfactory. It is of crucial importance that governments support the research that is needed to permit more precise predictions of the probable implications of change for regions and individual countries. Governments obviously vary greatly in their ability to support such activities unaided, but we believe that as a minimum individual countries should have, or should be helped to acquire, some capacity to assess their own climate and sea level, and changes in them, even if not all countries can contribute to the basic science. Without such science, and effective monitoring, much money may be wasted in inappropriate responses and much human anxiety risked by erroneous conclusions [5].

One of the most critical areas requiring research support is agriculture, especially in the arid and semi-arid areas. It is vital that research is carried out on potential improved varieties of millet, sorghum and cassava, which are key subsistence crops in these areas. The declaration from the 1990 Nairobi Conference on Global Warming and Climatic Change also laid emphasis on the urgent need for research into the world's grasslands, which is an important area of enquiry which has not received the attention it deserves.

> The role of the African grasslands in the global carbon cycle is not well understood. While forests have received considerable attention in climatic change discussions, African grasslands are an unknown quantity....The role of grasslands in the global climatic change equation needs urgent attention, and research programmes to generate reliable and long-term data need to be formulated and supported [6].

Jimmy Holmes/Panos Pictures

Solar power is probably the most promising of the alternative fuels, particularly for developing countries, but so far the cash spent on research is a fraction of that spent on nuclear research.

If developing countries are to adopt and apply appropriate policies in response to global warming, they must be informed about what is happening. They must be able to monitor any changes in their own climate and along their own coastlines, and to collaborate with neighbouring states in building up a regional picture. A high degree of scientific assistance and training as well as coordination of studies and results will have to be provided by the industrial world for the appropriate international organisations if these essential tasks are to be carried out.

Regional and national scenarios

At present, there is no reliable information on the likely effects of global warming at a regional or national level. It is, however, possible to examine the likely effects if certain plausible global warming scenarios are realised. This was done, for example, in both the United States and Australia where public officials examined the detailed implications for water supplies, public health, infrastructure and other impact areas if certain climatic changes were to occur as a result of global warming.

Such scenarios, or simulations, do not produce firm recommendations for action. Their value is that they help to highlight areas of particular vulnerability or ignorance. At the very least, they point the way towards areas where further research is needed. They can also help in planning; other things being equal, it is better to choose the option which offers the least risk if global warming occurs as predicted. Developing countries need financial and technical resources which will help them to create a range of scenarios for their own areas.

What kind of development?

The threat of global warming has stimulated the debate about what kind of development path the Third World should follow. Some environmental commentators feel that the global warming issue is yet another, and perhaps final, warning that industrial society is on a course towards self-destruction and, therefore, provides an example which should not be followed by the developing world. It is a view shared by a substantial number of Third World observers

Dumping in the development and environmental circles is associated with hazardous chemical waste or sub-standard products that are trucked to the South by the industrially developed North.

This form of physical dumping, though hazardous, is much easier to deal with than the unnoticed dumping in the South of ideas and concepts, hatched and bred in the North. To give but a few examples, infamous conceptual dumps in recent times include AIDS, Acid Rain, Ozone Depletion, Overpopulation and Global Warming, among others.

The environment and development scenes are characterised by new concepts and crises that emerge every five years. The new crisis almost immediately overshadows whatever environmental concerns there might have been.

The only consistent element about these so-called global concerns and crises is that most, if not all, emerge from the North. Africa receives them in carefully prepared information dossiers, scientific studies, seminars and global conferences.

We are never allowed time to even ponder over our own problems. As soon as we are about to appreciate the impact of desertification, we are told of the ozone layer depletion. Before we blink an eye, the entire globe is heating up and we are told to stop cooking . Finally, comes the AIDs phenomenon...

Do we stop worrying about availability of drinking water, food, soil depletion and basic energy needs so as to remain pondering about solutions to mirage-like crises that pop up every five years?

The most disturbing phenomenon is that with each global crisis dumped on us, there is a plethora of experts, consultants, and specialised NGOs that emerge from the blue ready to come and save Africa.

The conceptual dumps, and the responses that follow from the North, do more harm to the Third World environment and development policies than the actual problems. The South, and especially Africa, simply needs time to think!

Achoka Awori, Executive Director, Kenya Energy and Environment Organisation (KENGO), quoted in *Impact*, newsletter of Climate Network Africa, June 1991

who feel that the present model of development adopted by the majority of developing nations, in addition to being unsustainable, is a form of cultural and economic colonisation.

Many other developing country commentators, however, disagree. Certainly, politicians, and the populations which elect them, tend to conceive progress in terms which are largely those of the industrial nations. They want the possessions, comforts and conveniences of industrial society. The Chinese government, for

Sheila Gray/Format

Colliery waste polluting a beach in Britain. How far is the industrial world a warning to developing countries and how far a model to emulate?

example, is determined to provide every household with a refrigerator, an objective with which the population is likely to agree wholeheartedly.

There is a genuine fear among many Third World people that there is a hidden agenda in the global warming discussions: that the industrial countries would like to impose a form of development on the Third World which denies it the benefits industrial countries take for granted. The point is valid. It is not for the industrial world to use the threat of global warming as an excuse for trying to haul up the development ladder by which it has ascended to its present comfortable and dominant position. The debate about the development model it wants to follow is for each Third World country to resolve for itself.

This does not, however, mean that Third World countries which choose to follow the same path as the industrial countries need to make all the same mistakes. It is in everyone's interest that such development strategies should not be based on second-rate or outdated technology. Neither should any developing country feel that because of economic weakness it should adopt environmental standards which are no longer regarded as acceptable in the industrial world.

People in developing countries have to formulate their own particular developmental response, based upon their own needs, capacities and priorities, backed up by financial and technical resources from the industrial world. In this sense the global warming issue can play the role of catalyst in the development process. As the declaration of the Nairobi Conference stated:

Dealing with the long-term effects of climatic change will require Africa to develop flexible economic systems and institutions which can anticipate change and adapt relatively fast to emerging trends. Economic systems and institutions will therefore need to be innovative, flexible, adaptive and diversified. Enhancing the adaptive capability of the African countries will require significant increases in the application of science and technology to development and willingness to reform existing institutions to reflect the need for constant change. A higher premium will need to be placed on social learning [7].

Energy pricing for efficiency and equity

A crucial area where there is a total coincidence between the economic interests of the developing world and the reduction of the global warming threat, is in improving the efficiency of energy use.

Petroleum imports rank with food imports as the greatest drain on foreign currency earnings in a large number of Third World countries. Building electric power stations and distribution systems also consumes vast amounts of money and technical resources. Improved energy efficiency would ease the burden of these expenditures and allow scarce capital and foreign exchange resources to be devoted to other pressing needs.

Yet energy efficiency in developing countries is generally a great deal lower than in the industrial world. This is the result of a variety of factors, including old equipment, poor maintenance, and lack of the necessary skilled managerial and technical staff. In addition, most Third World governments do not have the financial resources to provide grants and other incentives to business, industrial and domestic users to encourage them to save energy. The box overleaf summarises a report on the constraints to electricity conservation in Brazil.

Governments do, however, have it within their power to set energy prices at a level which would encourage saving. This is particularly important in the electricity sector, which is the single most voracious consumer of capital investment in the developing world. In 1986, it was estimated that the total debts incurred by Third World electricity utilities were about US$180 billion, around a fifth of the total debts of the Third World. It is unlikely that the position is any better today.

Electricity is obviously necessary for development. But often its benefits flow primarily to richer people. It is the better-off urban householders and business people, rather than the poor, who use the greatest amounts of electricity. In the rural areas, it is the rich with the large land holdings, rather than poor subsistence farming families, who are connected to the electricity supply system.

Yet the electricity charges made by Third World electricity supply companies are often far below the costs of supply. The main reason for such low charges is the belief that they stimulate or help

Energy conservation: Brazil

A report on electricity conservation in Brazil pointed out that a variety of technical, economic, institutional and other factors inhibit the widespread adoption of conservation measures:

1) Gaps in the technology base: energy-efficient products are not always available and some energy-efficient items include imported components or materials, adding to the cost of the final product.

2) Subsidised electricity prices: many residential and commercial customers receive electricity at 50%-80% of the cost of supply which reduces their interest in conservation, and makes it hard for the utilities to build up funds for "non-essential activities" such as conservation programmes.

3) Economic instability: high interest rates and levels of inflation not only make it difficult to determine the payback on a conservation project but also make consumers want very rapid payback. High inflation also erodes electricity prices due to the time lag between consumption and bill paying.

4) Lack of information: architects, builders, domestic and industrial consumers lack accurate and accessible information about conservation measures and their cost effectiveness.

5) Split responsibilities: those purchasing the equipment (landlords, builders) often do not pay the running costs. Most public utilities are either responsible for power generation or its distribution, rarely for both. Distribution companies lose money in the short term through conservation and give only weak support to such ideas, yet they have the closest relationship with the consumers.

6) Sensitivity to first cost: Consumers or businesses may not be able to afford technology with higher initial costs but lower running costs. Such decisions are also tied in with lack of information about conservation options and lack of an economic incentive to conserve.

The report's recommendations included reducing sales and import taxes on energy-efficient technology; maintaining electricity tariffs at or above the average cost of service "rather than using them to combat inflation"; giving financial incentives to utility companies to promote and invest in conservation; encouraging a shift away from electricity-intensive industries; and adopting mimimum efficiency requirements for appliances.

Information from *Electricity Conservation in Brazil: Status Report and Analysis*, prepared for the University of São Paolo, Brazil, and the US Environmental Protection Agency, 1990

economic development. In practice, their main effect is to transfer resources from the poor to the rich. Allowing the poor access to electricity while pricing it in a way which makes it a more economically and environmentally sustainable option clearly requires a greater commitment among developing country governments to equitable electricity distribution. But industrial countries also have a vital role to play in implementing more energy-efficient policies in the South, according to a recent survey of energy and the environment:

> Compared with the industrial countries, the Third World produces only a tiny part of global pollution. It will grow larger. Sensible electricity pricing and a shift to gas will do little to slow that growth. One of the best things developed countries could do would be to encourage developing countries to adopt energy-efficient technologies. Well-directed aid is one way to achieve that. Far more important, though, will be to hasten the adoption of energy-efficient technologies in the industrial world. Most new inventions are made in the OECD area: from there they make their way south. If industrial countries are serious about pollution, they should first make sure their own energy prices capture the full weight of environmental costs. Then their companies, and companies in Third World countries, will find it pays to make products that use energy frugally [8].

Preventing deforestation

Some future scenarios for stabilising greenhouse gas emissions assume that deforestation in the Third World will be stopped. It would be highly desirable and most Third World governments would be glad if it happened. But there are many difficulties.

One reason for the loss of forests in the developing world is the expansion of subsistence agriculture. Timber loggers, beef ranchers and urban fuelwood suppliers also play a major part.

The pressure for more cultivable land comes partly from increasing population. But it is also because subsistence farmers are often working marginal land and do not have the resources required to increase their yields and keep their land productive. Instead, they are forced to farm a piece of land until its yields begin to decline, when they move on and try to find somewhere else to repeat the process.

Ron Giling/Panos Pictures

Tropical forest in Surinam. Biological diversity is beginning to be recognised as representing a potential economic resource.

It must also be recognised that the economic value and productive capacity of a standing forest, as currently computed, is extremely low. Even when people live there, obtaining fruit, nuts and other valuable products, this is a very low-intensity use of the land. Cutting the trees and turning the land over to farming is economically far more productive. There is, therefore, an almost irresistible economic pressure to get rid of forests and put the land to other uses, which for local people will almost invariably be agriculture. Logging is also a means of turning the low economic value of standing trees into valuable timber—a process which countries under great pressure to pay back foreign debt find it difficult to forego.

None of this is to suggest that the loss of tropical forests is anything but a tragedy for the world, both in terms of losing biological wealth and one of the primary sinks for carbon. But any search for lasting solutions must start from a realistic appraisal of the social and economic forces which are driving deforestation. Actions which do not take these factors into account are inevitably going to fail. In many countries it is not an overall shortage of land that is driving subsistence farmers into the forests, but that too few people own most of the land, so land reform is a key issue. In countries where many people are barely meeting their basic needs, preserving forest resources for the benefit of future generations may seem a luxury they can ill afford. If it is in the interest of all countries to conserve the world's forests, the international community should recognise this and compensate poorer countries accordingly.

Increasing fossil fuel consumption

The present population of the Third World is about 3.7 billion people, about three-quarters of the world total. They consume about 20% of the world's fossil fuels. If the gap in living standards is to be narrowed, even prevented from widening, an increase in the fossil fuel consumption of the developing world is inevitable.

This does not mean that fossil fuel consumption should be increased for its own sake. Neither does it mean that wasteful use of fossil fuel energy will cause development to take place, quite the contrary. It is simply that there is no available alternative fuel for the increased industrial production, electricity generation, transport and other productive activities which will have to take place if the living standards of the rural and urban people of the developing world are to rise significantly.

This is an issue which will have to be recognised and accepted in the area of energy assistance. There is already a growing reluctance on the part of some donor agencies to provide assistance to projects which involve fossil fuels—for example, supplying diesel pumps or diesel generators for a rural electrification project.

Renewable energy sources should, of course, be used when they are technically and economically suitable. But there are many occasions when this is not the case. Development projects which do not provide a positive economic rate of return do not promote economic development; they act as a drain on the economy of the developing country.

The fact is that development in the Third World is going to require a significant increase in fossil fuel consumption over the coming decades. Renewable energy sources will be able to make a limited contribution and should be used when suitable. But the main task is to ensure that the fossil fuel technologies used are economically justified and as energy-efficient as possible.

Debt, trade and aid

The resources available to Third World governments wishing to take action to avert the global warming threat are extremely limited in comparison with those at the disposal of the industrial nations. The full magnitude of the economic gulf between the Third World and the industrial countries is sometimes forgotten in discussions about global warming. But a few sample figures can help restore a sense of proportion.

The total economic output of France, with a population of 56 million, is about three times greater than that of India with its 900 million people. Norway, with just four million people, has an economic output about four times that of Bangladesh, with its 115 million people. The average Swedish GNP per person is 120 times greater than that of Tanzania.

The burden of debt is stifling. At present, the Third World is paying back about US$50 billion more to the industrial countries than it is receiving in new money each year. The net flow of investment capital is from Africa, Latin America and the Caribbean to Europe, the United States and Japan.

Much of the problem is a result of the huge budget deficits run by the United States during the 1980s. The need to borrow money to cover these deficits pushed up interest rates throughout the world. The devastating result for the Third World was that interest payments in the 1980s were five times higher than in the 1970s. In consequence, Third World countries have been selling food and consumer goods abroad; poverty alleviation programmes have been

---

Developing countries face particular problems in relation to climate change. They have an immediate, and immense, task in reducing poverty which climate change could make more difficult. To realise this goal they will need to achieve rapid rates of growth—albeit in a sustainable manner. They cannot, in these circumstances, be expected to curb that growth in order to alleviate a global problem which they have, in any event, done little to create. The burden of measures to reduce emissions will therefore fall overwhelmingly on the developed world.

"Climate Change: Meeting the Challenge", a report by a Commonwealth Group of Experts, Commonwealth Secretariat, London, 1989

stopped; investment has fallen; aging capital equipment has not been replaced; and economic growth has fallen.

The trade policies followed by the industrial world also create major problems for the developing nations. Tariff barriers prevent Third World goods from entering Northern markets lest they undercut domestic producers. At the same time, subsidised food exports are dumped in Third World countries undermining the markets for local farmers. As cigarette smoking declines in Europe for health reasons, the European Community even subsidises the export of tobacco.

As Vandana Shiva of the Research Foundation for Science and Ecology in Dehra Dun, India, has pointed out:

> Any solution to global warming will have to involve a long-term solution to the debt crisis. Debt is a major obstacle to the Third World conservation role that Third World citizens and the global ecological crisis are demanding. The Northern financial system is directly responsible for the debt crisis, through pushing bad loans and unjust interest demands on the Third World. As economist John Kenneth Galbraith has written: "Such loans, given by foolish banks to foolish governments for foolish purposes, generally are not—and perhaps should not be—repaid"[9].

Third World countries are as deeply involved in the problem of global warming as the industrial world. They need resources if they are to prepare themselves against its possible effects. They also need resources if they are to contribute to the fight against it. But at present they are so short of investment capital and so burdened with debt repayments that short-term survival is all that many can manage. If the world is to be mobilised for action against global warming, issues such as these will have to be addressed.

Appendix

International initiatives and organisations

A large number of organisations are now working on various aspects of the global warming issue. There is already a confusing proliferation of names and acronyms. The following notes briefly describe some of the main international organisations and the principal initiatives which have been taken.

The World Climate Programme

World Climate Programme, c/o Executive Secretary (Internegotiating Committee for a Framework Convention on Climate Change), Pavillons du Petit Saconnex, 16 Ave Jean Trembley, 1209 Geneva, Switzerland.

The First World Climate Conference was held in 1979 as a response to growing international unease about the potential impact of human activities on the global climate. One of the outcomes of the conference was the establishment of the World Climate Programme (WCP) under the auspices of the World Meteorological Organization (WMO), the United Nations Environment Programme (UNEP), the United Nations Educational, Scientific and Cultural Organization (UNESCO) and the International Council of Scientific Unions (ICSU).

WCP convened a workshop in 1985 at Villach, Austria, to review research on the greenhouse effect. A consensus was reached that the concentration of greenhouse gases in the atmosphere could double by around 2030, which could lead to a temperature rise of 1.5-4.5°C, although the full effect would not be felt for some decades.

Two further workshops were held in 1987 under the auspices of the Beijer Institute (now the Stockholm Environment Institute), one in Villach and the other in Bellagio, Italy. These confirmed the growing scientific consensus and drew up the agenda for an international conference to be held the following year in Toronto, Canada.

WCP activities also include a World Climate Data Programme which assists countries in setting up data collection systems; a series of research programmes into particular aspects of world climate; and climate impact studies of particular areas and different aspects of global warming.

The World Ocean Circulation Experiment (WOCE), also coordinated by WCP, is a huge scientific effort, involving 40 nations, to improve knowledge of the oceans' role in determining the global climate. It includes studies of ocean currents, surface and deep ocean temperatures, and wind patterns, using both satellites and survey ships.

The Second World Climate Conference

The Second World Climate Conference was held in late 1990 in Geneva, Switzerland. It was sponsored by WMO, UNEP, UNESCO and ICSU. Its discussions relied heavily on the outputs of the three IPCC Working Groups, with the warning that global warming was inevitable if the concentration of greenhouse gases continued to increase. Although the European Community nations agreed to stabilise greenhouse gas emissions by the year 2000, the United States and the USSR were adamant in resisting any specified targets.

The Advisory Group on Greenhouse Gases

Advisory Group on Greenhouse Gases, coordinated by the Stockholm Environment Institute, Box 2142, S-10314 Stockholm.

The Advisory Group on Greenhouse Gases (AGGG) was set up by WMO, UNEP and ICSU to ensure there was adequate follow-up of the 1985 WCP workshop in Villach. It compiles biennial reviews of international and regional studies related to greenhouse gases as well as periodic assessments of the rates of increase of greenhouse gas levels in the atmosphere.

The International Geosphere-Biosphere Programme

The International Geosphere-Biosphere Programme, Secretariat, Royal Swedish Academy of Sciences, Box 50005, S-10405 Stockholm, Sweden.

This programme was established by ICSU in 1986. Its purpose is to "describe and understand the interactive physical, chemical and biological processes that regulate the total earth system...the changes that are occurring in this system, and the manner in which they are influenced by human actions."

The programme builds on other work by ICSU, especially through its Scientific Committee on Problems of the Environment (SCOPE).

The 1988 Toronto Conference on The Changing Atmosphere

A conference entitled The Changing Atmosphere: Implications for Global Security was held in Toronto in June 1988. A total of 45 countries and 15 international organisations were represented.

The conference accepted the scientific evidence on global warming. Among its recommendations were the following:

- reduce carbon dioxide emissions by 20% of 1988 levels by the year 2005 through improved energy efficiency and modification of supply;

- halt deforestation and increase afforestation;

- strengthen the 1988 Montreal Protocol on the emissions of CFCs in order to eliminate emissions by the year 2000 and to reduce emissions of other ozone-depleting gases;

- initiate the development of a comprehensive global convention for protocols on the protection of the atmosphere;

- devote increased resources to research programmes concerned with scientific and policy aspects of the problems, to support the continuing assessment of research results and to stimulate governmental discussion of responses and strategies;

- establish a trust fund to assist Third World nations and to encourage these nations to participate in international efforts concerning monitoring, research, adaptation and control;

- increase funding to non-governmental organisations and educational establishments to permit the initiation and development of educational campaigns and programmes.

The Montreal Protocol

The lead role in coordinating international action to curb destruction of the ozone layer has been played by UNEP. This led to the Montreal Protocol on Substances which Deplete the Ozone Layer which was signed in 1988. It agreed a three-stage cut in the emissions of CFCs.

A number of major Third World manufacturing countries, including China, India and Brazil, were unwilling to sign the protocol unless they were given financial help and access to the technology for manufacturing CFC substitutes.

A further meeting in London in 1990 agreed to accelerate the process and phase out CFCs and other ozone-depleting gases by the year 2000. The new timetable envisages a 50% cut by 1995, an 85% cut by 1997 and final elimination by the year 2000.

Progress was also made at the 1990 meeting on the transfer of technology from the industrial world, with the Indian and Chinese delegates feeling able to recommend acceptance of the agreement to their governments. The Indian Environment Minister, Maneka Gandhi, said that the agreement now stipulated that if the technology for manufacturing substitutes was not transferred, there would be no obligation on India to stop making CFCs.

The 1992 United Nations Conference on Environment and Development

UNCED Secretariat, Case Postale 80, Conches, Switzerland.

The threat of global warming is an underlying issue for the United Nations Conference on Environment and Development (UNCED) to be held in Rio de Janeiro, Brazil, in 1992. A variety of organisations are preparing reports and action plans for presentation at the "Earth Summit".

An Inter-governmental Committee is negotiating a Framework Convention on Climate Change to be signed at UNCED.

World Commission on Environment and Development

The Centre for Our Common Future, Palais Wilson, 52 Rue des Pâquis, CH-1201 Geneva, Switzerland.

The World Commission on Environment and Development (The Brundtland Commission) was set up by the Secretary General of the

UN in 1983. It was chaired by Mrs Gro Harlem Brundtland, a former and subsequent Prime Minister of Norway. The Commission appointed a group of experts to assist its investigations; it also held public hearings in Indonesia, Brazil, Canada, Zimbabwe, Kenya, Moscow and Tokyo. Its report was presented to the UN General Assembly in late 1987 and published under the title *Our Common Future*. In 1988 the Centre for Our Common Future was established in Geneva to act as a focal point for all follow-up activities.

Regional conferences and NGO initiatives

There have been a number of conferences focusing on regional aspects of global warming. There has also been intensive NGO activity. Events are moving rapidly in this area, and this Appendix attempts only the broadest indication of what is happening.

Climate Action Network

CAN Europe, 98 Rue du Trone, Bte 8, 1050 Brussells, Belguim.
Climate Network Africa, PO Box 21136, Nairobi, Kenya.

The Climate Action Network (CAN) is an international information exchange between NGOs involved in global warming. It was founded at the 1988 Toronto Conference. It has a series of national and regional contacts throughout the world.

Climate Network Africa, which has links with CAN, is based in Nairobi. It is an informal coalition of African NGOs which aims to increase participation in national and international discussion on global warming.

New Delhi Conference on Global Warming and Climate Change

The Tata Energy Research Institute of New Delhi, India, together with the Woods Hole Research Institute, organised a conference entitled Global Warming and Climate Change: Perspectives from Developing Countries, in February 1989. The conference described global warming as "the greatest crisis ever faced collectively by humankind" and set out an agenda for action.

The Nairobi Declaration on Climatic Change

The African Centre for Technology Studies (ACTS) of Nairobi, Kenya, together with the Woods Hole Research Institute of Massachusetts, USA, organised a conference entitled Global Warming and Climatic Change: African Perspectives, in May 1990. The conference concluded: "There is sufficient scientific data to justify action being taken now to reduce greenhouse gas emissions and implement strategic planning to deal with the anticipated effects of climatic change." A 25-point plan of action for governments, international agencies and NGOs was issued.

Alliance of Small Island States

AOSIS, c/o Centre for International Environmental Law, Kings College, University of London, Manresa Road, London SW3 6LX, UK.

AOSIS is an alliance of 36 member states, mainly from the Pacific and Caribbean. It was formed at the Second World Climate Conference in late 1990. The main objective is to draw together the interests of the small island states, who are particularly at risk from climate change, and strengthen their voice in the international legal process and preparations for UNCED.

Notes

Chapter Three The Implications of Global Warming

1. "Climate Change: Meeting the Challenge", a report by a Commonwealth Group of Experts, Commonwealth Secretariat, London, 1989.

2. Ali, Sayed Iqbal and Huq, Saleemul, "International Sea Level Rise: A national assessment of effects and possible responses for Bangladesh", Bangladesh Centre for Advanced Studies, September 1990.

Chapter Four Reducing the Risks

1. Smith, Gar, "Australia Confronts Climate Change", *Earth Island Journal, Summer* 1991.

2. Reddy, Amulya, "Environmentally Sound Energy Development: A case study of electricity for Karnataka State", paper presented at the Regional Conference on Environmental Challenges for Asia- Pacific Energy Systems in the 1990s, Kuala Lumpur, Malaysia, January 1991.

3. *Energy for a Sustainable World*, Wiley Eastern Ltd, New Delhi, 1988.

Chapter Five Policy Implications for the Third World

1. *British Petroleum Statistical Review of World Energy*, 1991.

2. *Global Warming in an Unequal World: A case of environmental colonialism*, Centre for Science and Environment, New Delhi, 1991.

3. Ibid.

4. Agarwal, Anil, "UNCED in Perspective", *Costing the Earth: Striking the global bargain at the 1992 Earth Summit*, World Development Movement, London, 1991.

5. "Climate Change: Meeting the Challenge", op. cit.

6. "The Nairobi Declaration on Climatic Change", from the International Conference on Global Warming and Climatic Change: African Perspectives, published by the African Centre for Technology Studies, Nairobi, Kenya, 1990.

7. Ibid.

8. "Power for the poor", in "A Survey of Energy and the Environment", *The Economist*, London, August 1991.

9. Shiva, Vandana, "Growing addiction to fossil fuels", *Panoscope*, The Panos Institute, London, November 1990.

Index